Lecture Guide and Student Notes

CALCULUS

from Graphical, Numerical, and Symbolic Points of View

Ostebee and Zorn   **Volume 1**

Stephen M. Kokoska
Bloomsburg University

Saunders College Publishing

Harcourt Brace College Publishers

Fort Worth   Philadelphia   San Diego   New York   Orlando   Austin
San Antonio   Toronto   Montreal   London   Sydney   Tokyo

Kokosha; Lecture Guide and Student Notes to accompany
<u>Calculus from Graphical Numerical and Symbolic Points of View,
Vol I.</u> First Edition.

ISBN 0-03-017409-0

567 017 987654321

Preface

Technology, including computer software programs like *Mathematica*, *Maple*, and *Derive*, and graphing calculators, has significantly changed mathematics, particularly calculus. Students are now able to quickly visualize important concepts, explore possible solutions graphically, study and discover relationships among functions, and even check answers symbolically. Unfortunately, the teaching style used in many calculus classes has not changed but has remained a very traditional blackboard lecture-notetaking routine.

The purpose of this supplement is to ease the burden of taking notes, increase communication between instructor and students, and free class time for the use of technology. Typically an instructor copies theorems, definitions, and problems from the textbook to their notes, and then during class, to the blackboard. Students frantically copy the board, try to keep up the lecture, and worry about formula details. Questions from students are often in regards to penmanship and not about mathematics concepts. This traditional lecture style leaves little time for discussion of mathematics and technology demonstrations.

The *Lecture Guide and Student Notes* is an entire set of notes to accompany *Calculus from Graphical, Numerical, and Symbolic Points of View*, Volume 1, by Ostebee and Zorn. All of the Theorems, Definitions, and Facts are included so that neither student nor instructor has to copy them (before or) during class. There are plenty of examples with enough space left for solutions, and graphs with enough room for labels and notes. Instructors still retain enough flexibility to add their own examples and there is enough space on each page for students to add further explanations, comments that are often missed during a traditional lecture.

Instructor lectures may be based on overhead transparencies made from each page of this supplement. Since each student has the material already (their own LGSN), this serves as a starting point for discussions, examples, and justifications. This teaching method also permits an easy shift between transparency and technology and increases student participation. Students have time to ask and answer questions freely, and to grasp important concepts in class.

Over the past four years I have had success with this teaching style in a number of different courses. Students are pleased with the emphasis on mathematics versus taking notes. I seem to have more spirited discussions about mathematics during class, and it is much easier to work technology into the syllabus. The LGSN represents a very simple, efficient, alternative teaching style that makes it easier to *learn* rather than *copy* mathematics.

Stephen M. Kokoska

Contents

Chapter 1

Functions in Calculus

1.1 Functions, calculus-style

What is calculus?

(C1) The study of functions.

(C2) The study of (instantaneous) change.

> **Definition** (informal): A function is a procedure for assigning a unique output to any acceptable input.

Example: The most common functions are given explicitly by algebraic formulas.

$$f(x) = x^4 - 3x^2 + 15 \qquad g(x) = \frac{x-1}{x^2 - 2x - 3}$$ □

Note! Not all functions have a nice formula. A function may be described by a graph, table, words, etc. ◇

Example: The table below shows the amount of rainfall (in inches) from May to July and the number of bushels (in thousands) of corn harvested on a farm for the last twenty years.

rain	3.5	4.9	4.5	8.1	2.4	6.2	3.8	4.2	5.1	5.3
bushels	4.8	6.0	6.8	5.6	3.5	7.0	6.1	7.5	7.5	6.2
rain	6.5	5.2	3.0	2.2	6.5	1.5	4.6	5.1	9.6	5.8
bushels	4.5	8.9	3.0	2.8	6.3	2.1	7.8	8.7	2.6	7.6

Here is a plot of the data.

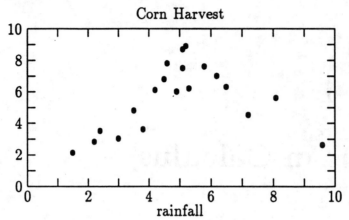

The graph (called a scatterplot) seems to indicate there is some "optimal" amount of rainfall.

It may be possible to *fit* the data with a reasonable function. The picture below shows the graph of a function c, given by $c(r) = -.32r^2 + 3.68r - 3.64$ superimposed on the data.

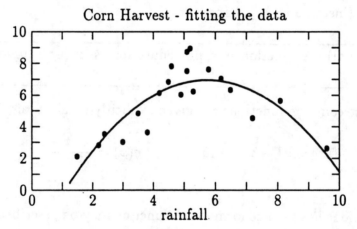

Note!

(N1) **parameters**: various constants in the formula.
curve fitting: finding appropriate parameters.

(N2) Might expect c to be *continuous*, or to vary smoothly.

(N3) c might not be given *exactly* by a simple algebraic formula, valid for any amount of rainfall. But we might try to model or estimate c.

(N4) Most real-world functions are messy, and do not have an explicit formula.
We can still be successful at modeling and prediction.   ◇ □

Example: Let n be the function shown graphically below.

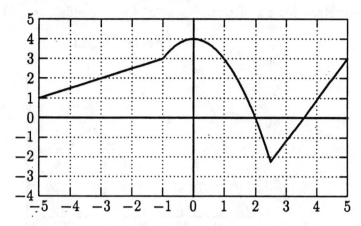

Some values of n are easy to compute, some aren't.

$n(-3) =$

$n(1.5) =$

n can be represented by a piecewise-defined formula:

$$n(x) = \begin{cases} \dfrac{1}{2}x + \dfrac{7}{2} & \text{if } -5 \leq x \leq -1 \\[2mm] 4 - x^2 & \text{if } -1 < x < 2.5 \\[2mm] 2.1x - 7.5 & \text{if } 2.5 \leq x \leq 5 \end{cases}$$ □

Example: Let $V(x)$ denote the volume (in gallons) of water in a spherical tank, of radius 10 feet, when the depth is x feet.

(E1) What volume of water is in the tank when the water is 10 feet deep?

(E2) Draw a rough graph of the function $V(x)$ for $0 \leq x \leq 20$.

□

Example: The function h (shown below) returns the height (in feet) above the ground of a roller coaster over a certain period of time (measured in seconds).

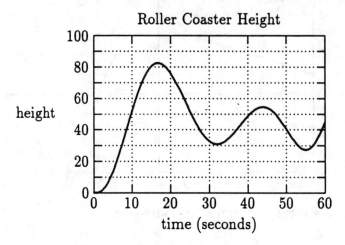

The graph tells us a lot. Here are some ideas to think about.

(E1) What is the height of the roller coaster at any time t?

(E2) How fast was the roller coaster rising or falling at time t?
      What is the *upward velocity, $V(t)$*?

      Note!

      (N1) If $V(t) > 0$ then the roller coaster is rising.
            If $V(t) < 0$ then the roller coaster is falling.

      (N2) At $t \approx 16$ the height levels off, $V \approx 0$.   ◇

(E3) $V(t)$ depends upon how steep the graph of h is!

(E4) How would you estimate $V(10)$? How about $V(11)$?

(E5) How would you graph the function V? (See the figure below.)

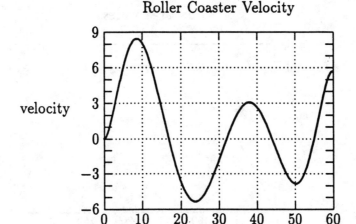

Roller Coaster Velocity

$\square$

Remarks:

(R1) Mathematics is a language. It stresses simplicity, is straightforward, and precise.

(R2) Calculus is a language used to describe changing quantities. Calculus also provides us with a set of rules and regulations used to predict change. $\triangle$

Example: A farmer has 700 feet of fencing and he wants to enclose a rectangular region along a river. He does not need any fencing along the river. What are the dimensions of the field that has the largest area?

Here is a diagram and some notation to help us with this problem.

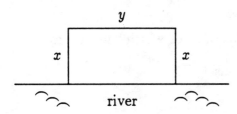

Let A be the area of the field.

For any given x, A can be found algebraically:

$$2x + y = 700$$

$$y = 700 - 2x$$

$$A(x) = x \cdot y = x(700 - 2x) = 700x - 2x^2$$

A graph shows how A varies with x.

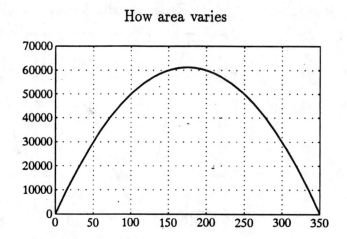

If we "zoom in" we get a better estimate.

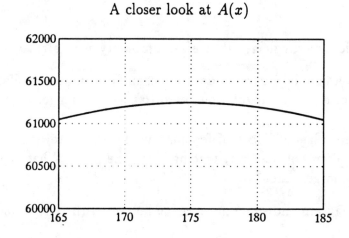

We can estimate the maximum area, and the value of x for which this occurs.

Calculus allows us to find these answers more precisely.   □

Remarks:

(R1) It is common practice in calculus to put some *old* functions together in order to create a new function.

(R2) Often, we also try to approximate one function with another. △

Example: Consider the function $p(x) = x^2$. Define a new function, D, based on p, to be the distance from the point $(4, -1/2)$ to the point $(x, p(x))$ on the curve p. Find a formula for $D(x)$; draw the graph.

Points on p have the form $(x, p(x)) = (x, x^2)$.

Using the distance formula:

$$D(x) = \sqrt{(x - 4)^2 + (x^2 - (-1/2))^2} = \sqrt{(x - 4)^2 + (x^2 + 1/2)^2}$$

Here is a graph of $y = D(x)$.

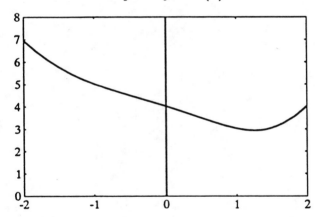

Graph of $y = D(x)$

There seems to be a *low* spot on D, and it falls somewhere around $x = 1$. This spot is the point on p that is closest to $(4, 1/2)$. □

1.2   Graphs

Definition: The **graph of an equation** in x and y is the set of ordered pairs (x, y) of real numbers that satisfy the equation.

Remarks: Try graphing these three equations:

$$x^2 + y^2 = 2xy ; \qquad x^2 + y^2 = 2xy + 4 ; \qquad x^2 + y^2 = 2xy - 4 \qquad \triangle$$

Definition: The **graph of a function** f is the set of points (x, y) that satisfy the equation $y = f(x)$.

Remarks:

(R1) In other words: The graph of a function is the set of all points of the form $(x, f(x))$.

(R2) Here is a generic picture.

(R3) It is easy to draw graphs now using calculators and computers. It is still important to understand and interpret graphs.

(R4) Not every function has a nice formula. $\qquad\qquad \triangle$

Example: Consider the graph of $p(x) = x^3 - 3x + 1$.

Graph of $y = p(x)$

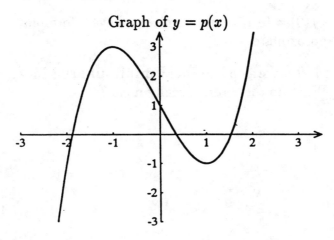

y-intercept: The graph intercepts the *y*-axis at 1. $p(0) = 1$.

No breaks: The graph can be drawn "without lifting your pencil." The function *p* is continuous.

x intercept: Solve the equation $p(x) = 0$. $x \approx -1.88, .35, 1.53$

When is *p* increasing? decreasing?

□

Example: Here is an example of a graph of a function *f*. No formula for *f* is given.

Graph of $y = f(x)$

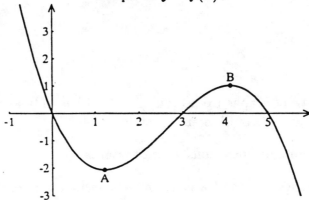

The graph can tell us a lot about the function f.

It sure looks like $f(0) = 0$. It would be nice to have a formula to check this, graphical data is always approximate.

The points A and B correspond to **local minimum** and **local maximum** values of the function f. What do these terms really mean?

Look at **concavity**. Where is the graph **concave up**? **concave down**? Where does it change concavity (**inflection point**)?

□

Remarks:

(R1) We can also construct graphs from tabular data. If we only know a few points on the graph of f, there are lots of different possible graphs.

(R2) New functions from old: operations with constants.

Given an "old" function f and a constant a, consider the following "new" functions and their effect on the graph of f.

(N1) $f(x) + a$: shifts the graph of f a units upward, a vertical translation.

(N2) $f(x + a)$: shifts the graph of f a units left, a horizontal translation.

(N3) $af(x)$: stretches or compresses the graph of f vertically by a factor of $|a|$, if $a < 0$ then also a reflection about the x-axis.

(N4) $f(ax)$: stretches or compresses the graph of f horizontally, if $a < 0$ then also a reflection about the y-axis.

◇ △

1.3 Machine graphics

- Calculators and computers allow us to represent objects geometrically, to quickly graph a function.

- Consider some common terminology.

- Computer graphics can be misleading!

Remarks:

(R1) When we graph a function f, we really only see *part* of the graph of f. Partial pictures are all we can get. The term *complete* graph is sometimes used to mean all of the important features of the graph.

(R2) The **viewing window** describes what *is* shown. The viewing window is a rectangle and can be represented in product notation.

$$\{(x,y)| -8 \leq x \leq 8, \; -3 \leq y \leq 3\} = [-8,8] \times [-3,3]$$

Note!

(N1) The two intervals are sometimes called the xrange and yrange. Together they are called plotting parameters.

(N2) The viewing window does not have to be symmetric about 0.

(N3) The appearance of the graph depends upon the viewing window.

Here is an example of the graph of a function f drawn in 6 different viewing windows.

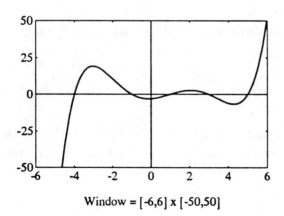

Window = [-6,6] x [-50,50]

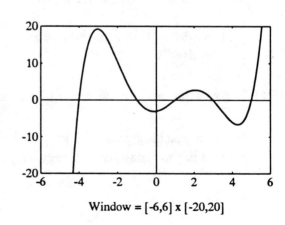

Window = [-6,6] x [-20,20]

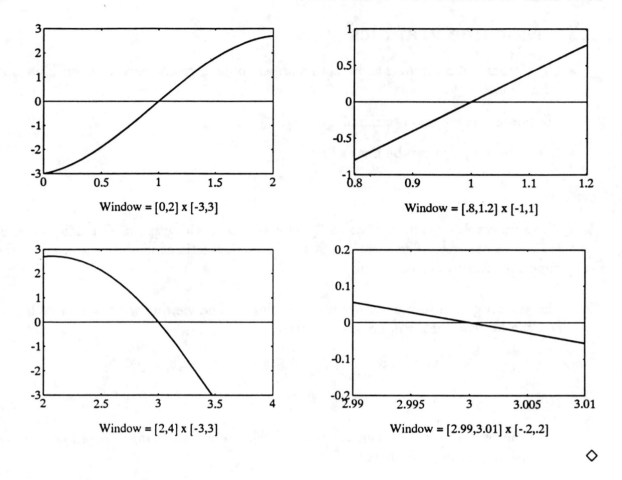

(R3) A little plotting jargon is in order.

(J1) Scale: The size of the horizontal or vertical units on a graph.

(J2) Zooming in: Enlarging the scale, by a zoom factor, in one or both axis directions.

(J3) Zooming out: Shrinking the scale in one or both axis directions.

(J4) Aspect ratio: The relative sizes of horizontal and vertical units. (Why would this matter?)

(R4) Distrust large windows and ill-behaved functions.

When a machine plots the graph of a function, it "connects the dots." If the plotting sample is too small, or too regularly spaced the function's behavior may be misrepresented.

(R5) Cautions: machine graphics should be viewed with healthy skepticism. Sometimes, not everything is as it seems.

(C1) Too few data: odd shaped graphs may require a smaller window, lots of data points.

(C2) Awkward windows: Let $f(x) = (x-1)^2(x+1)^2x^4$. Graph details may be lost depending on the window. Consider the graph of f in the two windows below.

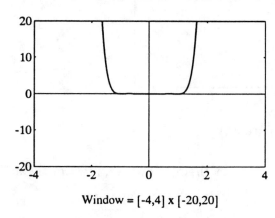 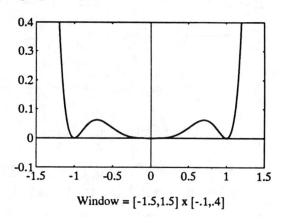

Window = [-4,4] x [-20,20]   Window = [-1.5,1.5] x [-.1,.4]

(C3) Scale effects: different scales may show different aspects of the function's behavior.

(C4) Slippery slopes: Let $f(x) = \sin x$ and $g(x) = 10\sin x$. The graph of g should be steeper than the graph of f (as g oscillates between -10 and 10, f varies only between -1 and 1). But the graphs below look almost the same!

$f(x) = \sin x$ $g(x) = 10\sin x$

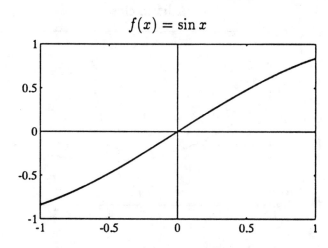 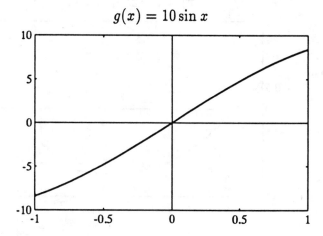

Now consider both graphs in the same window.

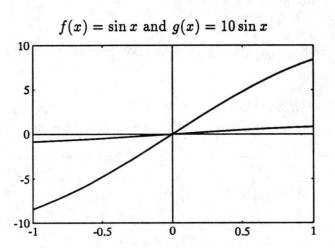

$$f(x) = \sin x \text{ and } g(x) = 10 \sin x$$

Watch vertical and horizontal scales carefully.

(R6) We will look at a lot of graphs, and try to interpret them. However, most of the actual
drawing we will leave to the calculator or computer.   △

Example: Find the minimum value, and when it occurs, of the function $f(x) = \cos 2x + 2 \cos x$
over the interval $[0, 3]$.

Pretty soon we will be able to use calculus to solve this problem. For now, consider a
few graphs.

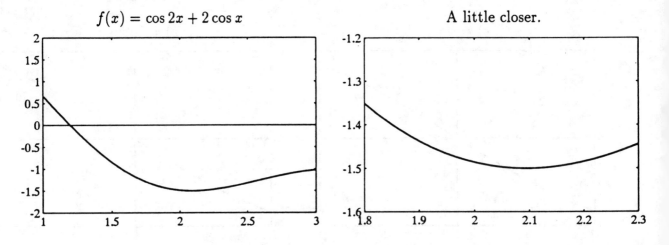

$f(x) = \cos 2x + 2 \cos x$ A little closer.

And still closer.

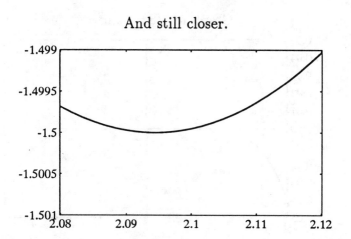

Conclusion: The minimum value is about -1.5 and it occurs near $x = 2.09$.   □

Example: For x near 0, the function $p(x) = x^2 - x^4/3$ is a pretty good approximation to $\sin^2 x$ (we'll learn how to find functions like p later). How well does p approximate $\sin^2 x$ on $[-1.5, 1.5]$? For which values of x can we be sure that $p(x)$ differs from $\sin^2 x$ by less than .001?

First, compare the two graphs.

Graphs of $y = \sin^2 x$ and $y = x^2 - x^4/3$

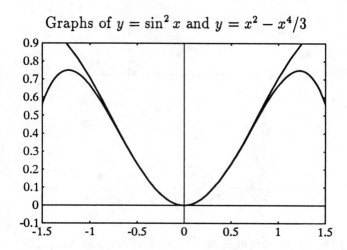

For $-1 \le x \le 1$ the approximation looks good. Outside this interval, p is off the mark.

On $[-1, 1]$ consider the error function $p(x) - \sin^2 x$.

Graph of $p(x) - \sin^2 x$

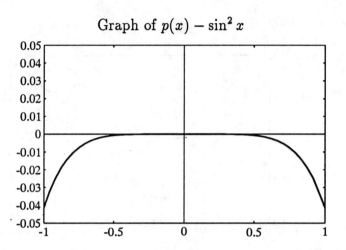

On $[-1, 1]$, $p(x)$ approximates $\sin^2 x$ with error no worse than .05 (maybe even a little better than that).

For what values of x is $|p(x) - \sin^2 x| < .001$?

Take a look at another graph.

Graph of $p(x) - \sin^2 x$:   a closer look

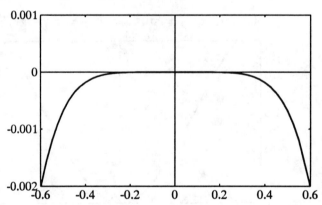

Conclusion: $p(x)$ approximates $\sin^2 x$ with error less than .001 if $|x| < .55$.   □

1.4   What *is* a function?

- We have looked at functions in earlier sections, defined by formulas, by words, by graphs, and even tables.

- Now, we will define some terms more carefully and precisely.

Definition: A function is a **rule** for assigning to each member of one set, called the **domain**, *one* member of another set, called the **range**.

Note!

(N1) Rule: A function accepts an input and assigns an output. The rule describes how this assignment is made.

(N2) Domain: Denoted $\mathcal{D}$, is the set of acceptable inputs.

(N3) Natural domain: the set of inputs for which the function makes sense.

(N4) Range: The set of outputs. If f has domain $\mathcal{D}$, then the range of f, denoted $\mathcal{R}$, is the set $\{f(x) \mid x \in \mathcal{D}\}$. (The set of all images of x under f) $\Diamond$

Example: Let $f(x) = \dfrac{x}{x^2 - 1}$. Discuss the rule, domain, and range associated with the function f.

> *Note*! The particular letter used to define a function doesn't matter. The following rules all define exactly the same function.

$$f(x) = \frac{x}{x^2 - 1}\,; \qquad f(t) = \frac{t}{t^2 - 1}\,; \qquad f(u) = \frac{u}{u^2 - 1} \qquad \Diamond$$

Example: Let $F(t)$ = the temperature in Trenton, NJ at any time t. Discuss the rule, domain, and range associated with the function F.

> *Note*! There is no convenient formula for F. But F makes sense as a function, it *describes* outputs, and at any one time, there is only one temperature.   ◇

□

Example: Consider $g(t) = \left\{ \begin{array}{rl} t+3 & \text{if } t < -1 \\ t^2 - 2t & \text{if } -1 \leq t < 3 \\ 4 & \text{if } t \geq 3 \end{array} \right\}$.

Discuss the rule, domain, and range associated with the function g.

> *Note*! g is defined by *cases*, and is called a **piecewise-defined** or **multiline function**. This kind of a function arises naturally in many applications.   ◇

□

Example: Let h be the function given by the graph below.

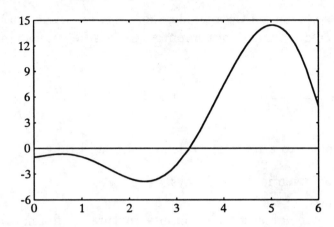

Discuss the rule, domain, and range associated with the function h.

$\square$

Example: Let j be the function given by the table below.

u	-2	-1	0	2	4	5	π	6.5
$j(u)$	6	5	-1	-1	3	7	9	8

Discuss the rule, domain, and range associated with the function j.

$\square$

Remarks:

(R1) How does technology handle the domain and range of a function? Try graphing these functions with a calculator and/or computer. Make sure you can interpret the resulting graphs.

$$f(x) = \frac{1}{(x-2)^2}\ ; \qquad g(x) = \frac{x+3}{x^2-9}\ ; \qquad h(x) = \sqrt{x^2-4}$$

(R2) There are *three* parts to a function: rule, domain, and range.

    (P1) Rule: may be given in various ways.

    (P2) Domain: the set of acceptable inputs (natural domain: all inputs that make sense).

    (P3) Range: set of output values.   △

Definition: A function f is **even** if $f(-x) = f(x)$ for all x in its domain; f is **odd** if $f(-x) = -f(x)$.

Note!

(N1) The graph of an even function is symmetric about the y-axis.

(N2) The graph of an odd function is symmetric about the origin.   ◇

Example: Consider the function $f(x) = x^4 - 3x^2$.

(E1) Show f is an even function.

(E2) Sketch the graph of f. Note the symmetry.

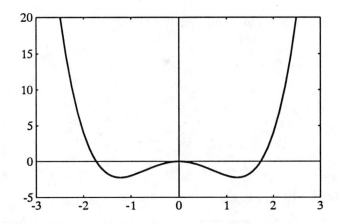

Example: Consider the function $g(x) = 4(x^3 - 3x)$.

(E1) Show g is an odd function.

(E2) Sketch the graph of g. Note the symmetry.

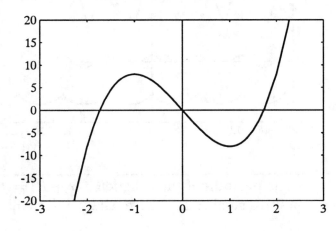

Example: Consider the function $h(x) = 3x - x^2$.

(E1) Determine whether h is even, odd, or neither.

(E2) Sketch the graph of h.

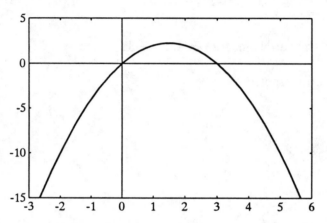

Informally, a function is **periodic** if it repeats itself on intervals of any *fixed* length. The basic trigonometric functions are periodic. Here are the graphs of three periodic functions.

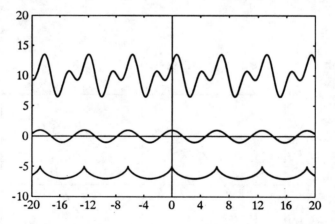

Definition: A function f is **periodic** if the equation $f(x + P) = f(x)$ holds for all x in the domain of f and a positive number P. The smallest such number P is the **period of** f.

Note!

(N1) Suppose f is periodic with period 2. Then $f(x+2) = f(x)$. This means that for any x the height of the graph at x is the same as the height of the graph at $x+2$.

(N2) For any function f, the graph of $y = f(x+P)$ is a horizontal shift of the graph of $y = f(x)$ p units to the left. So, if f is periodic with period P, the graph of f doesn't change if it is shifted P units to the left.

(N3) Technically, there are many periods of a periodic function. We usually only want the smallest period. ◇

Remarks (concerning functions):

(R1) It is important to understand the difference between a *function* and the *value of a function*.

f : a function; $f(x)$: a value of f

It is correct to say: Consider the function f defined by $f(x) = \sin x$.

Often we leave out a part of this sentence: Consider the function $f(x) = \sin x$.

This isn't technically correct, but we'll have to accept it and understand the proper use of these terms.

(R2) Composite expressions are tricky.

Consider the function f defined by $f(x) = x^2 + 3x$.

Find $f(x+h)$, $f(-x)$, $f(f(x))$

(R3) Be careful to distinguish between a *function* and an *expression*. They differ in subtle ways.

(S1) Not all functions are defined by symbolic expressions.

(S2) Several different symbolic expressions may define the same function.

(S3) Expressions with different variables may define the same function.

$\triangle$

1.5 A field guide to elementary functions

An **elementary function** is one built from certain legal basic *elements*, using certain legal operations.

Here is an example: $f(x) = \left(\dfrac{\cos(x^2)}{\ln(1+x)} \right)^4 + e^{\sin(x)}$

Note!

(N1) Basic elements: powers of x, trigonometric functions, log functions, etc.

(N2) Legal operations: $+$, $*$, $/$, $\circ$ (composition).

(N3) Elementary functions can get pretty complicated. ◇

Function Families:

(F1) Algebraic functions: Their rules involve only algebraic operations.

$$f(x) = x^2 + 4x + \frac{3}{x} \, ; \qquad g(x) = \frac{3x+7}{x^3 + 3x + 1}$$

(F2) Exponential and logarithm functions: Their rules involve exponentials or logarithms with various bases.

$$f(x) = 3^x \, ; \qquad g(x) = e^{2x} \, ; \qquad h(x) = \log_8 x \, ; \qquad k(x) = \ln x$$

(F3) Trigonometric functions: The sine, cosine, tangent, secant, cosecant, and cotangent functions.

Note!

(N1) There are *hybrid* functions, for example, $f(x) = \sin x + e^x$.

Here, functions from two different families combine to form a new function.

(N2) We'll start by looking at the basic building-block functions: identification, classification, and behavior. ◇

Algebraic functions

An **algebraic function** is one defined using only the ordinary algebraic operations: addition, multiplication, division, raising to powers, and taking roots.

Definition: A **polynomial** in x is an expression that can be written in the form

$$a_0 + a_1 x + a_2 x^2 + a_3 x^3 + \ldots + a_n x^n,$$

that is, a sum of *constant* multiples of *nonnegative integer* powers of x.

Example: Here are some examples of polynomials.

$$x^3 + 3x^2 + x + 5; \qquad -x^4 + x^{23}; \qquad (x+7)(ex^5 + 42x^4)$$

The following are *not* polynomials. Why not?

$$\sqrt{x}; \qquad x^3 + 3x^2 + 3x + 1 + \frac{1}{x}; \qquad \frac{x^3 + 3x + 1}{x^4 - 3x - 5} \qquad\qquad \square$$

Remarks:

(R1) The polynomial expression $x^4 + x^2 + 1$ corresponds naturally to the function $p(x) = x^4 + x^2 + 1$. Any function defined by a polynomial expression is called a **polynomial function**. The terms *polynomial* and *polynomial function* are used interchangeably.

(R2) Every polynomial function has the same natural domain: all real numbers.

(R3) Polynomials' ranges vary a lot. Consider $f(x) = 4 - x^2/3$, $g(x) = x^3 - 5x + 1$, and $h(x) = -4$. The graphs of all three are given below; which is which? And what is the range of each polynomial?

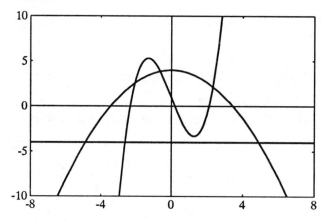

(R4) The term with the highest power of x determines the long run behavior of a polynomial, that is, what happens as $x \to +\infty$ and as $x \to -\infty$.

Consider the functions f, g, and h from (R3). What is the long run behavior of each of these functions?

(R5) We have already learned that the appearance of any graph depends upon the viewing window. Here is another look at f, g, and h in different viewing windows.

Graphs of f, g, and h: a closer view   Graphs of f, g, and h: a farther view

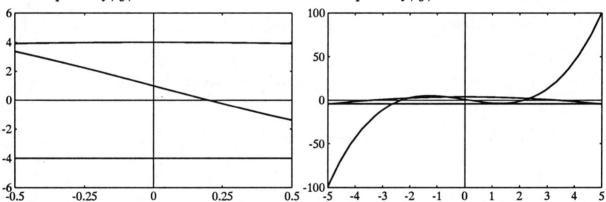

(R6) Polynomials do not have any *breaks* or *corners*. They are *continuous* and *smooth* everywhere. △

Special polynomials:

(S1) Constant functions: for example, $p(x) = -6$.

(S2) Linear functions: polynomials that involve nothing higher than the first power of x.

Some examples:

$$\ell_1(x) = x - 5; \qquad \ell_2(x) = x/2; \qquad \ell_3(x) = -3x + 4; \qquad \ell_4(x) = -4x - 5$$

Graphs of linear functions are lines. Here are the graphs of ℓ_1, ℓ_2, ℓ_3, and ℓ_4. Which is which?

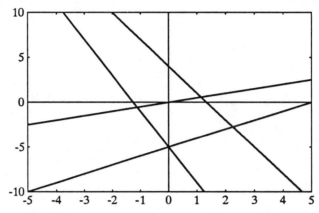

Here are some important terms.

(T1) Slope: the ratio $\dfrac{y_2 - y_1}{x_2 - x_1}$ where (x_1, y_1) and (x_2, y_2) are any two points on the line.

(T2) Lines with equal slopes are *parallel*. Horizontal lines have slope 0. Vertical lines have no slope, or undefined slope.

(T3) Point-slope form: The line through (a, b) with slope m has equation
$y = m(x - a) + b$.

(T4) Slope-intercept form: The line with slope m and y-intercept b has equation
$y = mx + b$.

(S3) Quadratics: polynomial of degree 2. Graphs of quadratic polynomials are parabolas. Here are a few examples.

$$q_1(x) = x^2 + 2x + 4; \qquad q_2(x) = (x - 3)(x - 4); \qquad q_3(x) = 5 - x^2$$

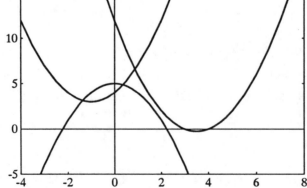

(S4) Cubics: polynomial of degree 3. Graphs are called cubic curves. Here are a few examples.

$$c_1(x) = x^3 - 3x - 3; \qquad c_2(x) = (1/2)(x-3)(x+4)(x-1); \qquad c_3(x) = -3x^3 + 4$$

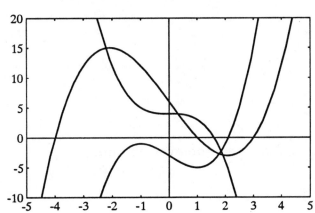

Note! Polynomials can take on almost any smooth, unbroken shape. However, a polynomial of degree n can have at most n roots, or can *turn around* at most $n-1$ times. ◇

Definition: A <u>rational function</u> is a function whose rule can be written as the quotient of two polynomials.

Example: Here are some rational functions.

$$f(x) = \frac{3}{x}; \quad g(x) = \frac{1-4x}{2x+2}; \quad h(x) = \frac{1-x^2}{x-2}; \quad j(x) = \frac{x^2+2x+1}{x^2-4} \qquad \square$$

Note! The graph of a rational function may contain horizontal and/or vertical asymptotes.

Consider $j(x) = \dfrac{x^2 + 2x + 1}{x^2 - 4}$ and its graph.

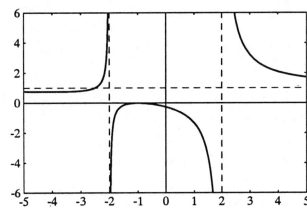

An asymptote is a straight line toward which the graph *tends*. Horizontal asymptotes reflect long run behavior.

What are the asymptotes for the function j?

Where do the vertical asymptotes come from?

$\diamond$

Note! There are other algebraic functions that involve n^{th} roots. Here are some examples.

$$f(x) = \sqrt{x+3}; \qquad g(x) = \sqrt{\sqrt{x-4}}; \qquad h(x) = \sqrt[3]{x + x^2 + \sqrt{x}}$$

Note the difference between the *exact value* and the *approximation*.

$$f(14) = \sqrt{17} \approx 4.1231 \qquad\qquad \diamond$$

Exponential and logarithm functions

These functions are very important in calculus, and practical in the real-world.

Facts:

(F1) Logs and exponentials are related.

For a positive base b and real numbers x and y,

$$y = \log_b x \iff x = b^y$$

(F2) Properties:

(P1) $b^x b^y = b^{x+y}$; $\qquad b > 0,\ x, y$ reals

(P2) $(b^x)^r = b^{xr}$; $\qquad b > 0,\ x, y$ reals

(P3) $\log_b xy = \log_b x + \log_b y$; $\qquad b > 0,\ x, y > 0$

(P4) $\log_b(x^r) = r \log_b x$; $\qquad b > 0,\ x, y > 0$

Definition: An **exponential function** is defined by an expression of the form $f(x) = b^x$, where b, the **base**, is a fixed positive number.

Remarks:

(R1) Here are some exponential functions.

$$3^x; \qquad 5.5^t; \qquad \left(\frac{2}{5}\right)^w; \qquad e^x$$

The following are *not* exponential functions.

$$w^2; \qquad (t+1)^5; \qquad (y^2+4)^5$$

(R2) The most important exponential function is the one with base e.

$e = 2.71827182\ldots$ $\quad$ (irrational)

$f(x) = e^x = \exp(x) = \exp x$: the (natural) exponential function.

(R3) Consider the graphs of $y = 1^x$, $y = (1/3)^x$, $y = 2^x$, $y = e^x$, $y = 5^x$ given below. (Which is which?)

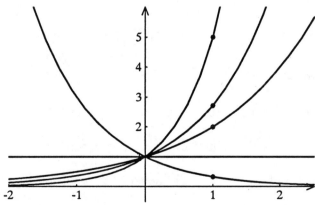

(G1) Any base $b > 0$ defines a sensible exponential function. ($b = 2, e, 10$ are the most common.)

(G2) Domain of all exponential functions: $(-\infty, \infty)$

Range of all exponential functions (unless $b = 1$): $(0, \infty)$

(G3) The larger the value of b, the faster b^x increases.

If $b < 1$, then b^x decreases. If $b = 1$, then b^x is a constant function.

(G4) Every graph of the form $y = b^x$ passes through the point $(0, 1)$.

And, if $f(x) = b^x$, then $f(1) = b^1 = b$

(G5) All exponential functions $f(x) = b^x$ are monotonic, either nonincreasing everywhere or nondecreasing everywhere.

If $b > 1$: always increasing;   if $b < 1$: always decreasing.

(G6) If $y = b^x$, then the rate of change of y is proportional to y itself. $\triangle$

Note! Exponential and power functions are very different. Consider the graphs of $y = x^3$ and $y = 3^x$.

(N1) The graph of $y = x^3$ has no asymptotes.

The graph of $y = 3^x$ has a horizontal asymptote. What is it?

(N2) Both x^3 and $3^x \to \infty$ as $x \to \infty$. But, 3^x grows *much* faster than x^3. $\diamond$

Logarithm functions

(L1) Exponential and logarithm functions come in pairs.

(L2) Whatever the exponential function does, the logarithm function undoes.

(L3) The logarithm function is really defined in terms of an exponential function.

Definition: The **logarithm function with base** b, denoted $f(x) = \log_b x$, is defined by the condition

$$y = \log_b x \iff b^y = x.$$

Remarks:

(R1) Here are some logarithm functions.

$$\log_3 x; \qquad \log_{1/3} x; \qquad \log_7 x; \qquad \log_e x$$

(R2) $\log_e x = \ln(x) = \ln x$: the natural logarithm function.

(R3) Consider the graphs of $y = \log_{1/3} x$, $y = \log_2 x$, $y = \log_e x$, $y = \log_3 x$, and $y = \log_{10} x$ given below. (Which is which?)

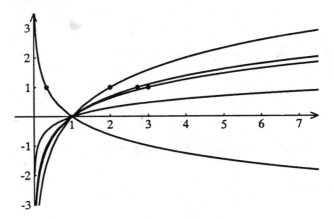

(G1) Any positive number b except $b = 1$ can be used as a base for a logarithm function. ($b = 2, e, 10$ are most common.)

(G2) Domain of every logarithm function: $(0, \infty)$

Range of every logarithm function: $(-\infty, \infty)$

(G3) The larger the value of b, the more slowly $\log_b x$ increases.

If $b < 1$, then $\log_b x$ decreases.

(G4) The graph of every logarithm function passes through the point $(1, 0)$

And, if $f(x) = \log_b x$, then $f(1) = \log_b 1 = 0$

(G5) Every logarithm function is strictly monotonic.   △

Note! Logarithm and exponential functions are inverses. Here is a graph of $y = 3^x$ and $y = \log_3 x$ to illustrate this property.

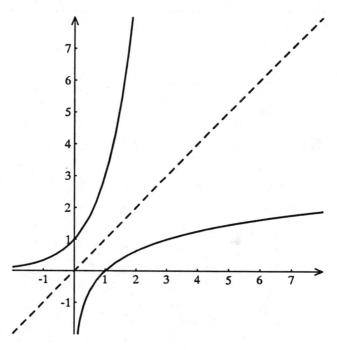

The graph of one is the reflection of the other across the line $y = x$.   ◇

Trigonometric functions

(T1) Trigonometric functions are periodic.

(T2) Here is a look at the graphs of the sine and cosine.

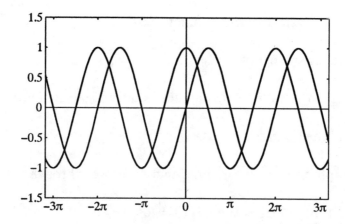

(T3) The definitions of the sine and the cosine function come from the unit circle.

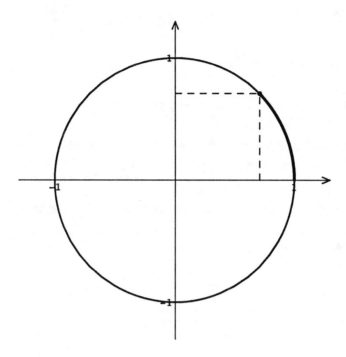

Definition: For any real number x, let $P(x)$ be the point reached by moving x units of distance counterclockwise around the unit circle, starting from $(1,0)$. (If $x < 0$, go clockwise.) Then

$$\cos(x) = u\text{-coordinate of } P(x)\,; \qquad \sin(x) = v\text{-coordinate of } P(x).$$

Remarks: about the sine and cosine function.

(R1) Domain: $(-\infty, \infty)$ for both sine and cosine.

Range: $[-1, 1]$ also for both.

(R2) You should *know* $\sin x$ and $\cos x$ for some common values of x.

Some common values: $0, \ \pi/6, \ \pi/4, \ \pi/3, \ \pi/2, \ldots$

(R3) $\sin(-x) = -\sin(x)$ sine function is odd.

$\cos(-x) = \cos(x)$ cosine function is even.

(R4) Sine and cosine functions are 2π-periodic.

$$\sin(x + 2\pi) = \sin(x) \qquad \text{and} \qquad \cos(x + 2\pi) = \cos(x)$$

(R5) Trigonometric identity: $\sin^2 x + \cos^2 x = 1$

There are lots of other identities: see Appendix G.

(R6) Trigonometric functions are transcendental. They cannot be written as algebraic combinations of powers, roots, sums, products, etc.   △

Four other trigonometric functions are defined in terms of the sine and cosine functions.

Definition: For real numbers x,

$$\tan x = \frac{\sin x}{\cos x} \qquad \cot x = \frac{\cos x}{\sin x} \qquad \sec x = \frac{1}{\cos x} \qquad \csc x = \frac{1}{\sin x}$$

Here are some graphs:

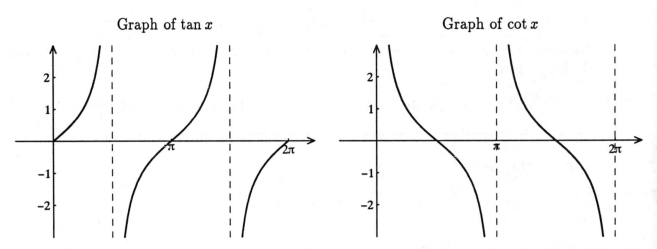

Graph of $\tan x$ Graph of $\cot x$

Graph of sec x Graph of csc x

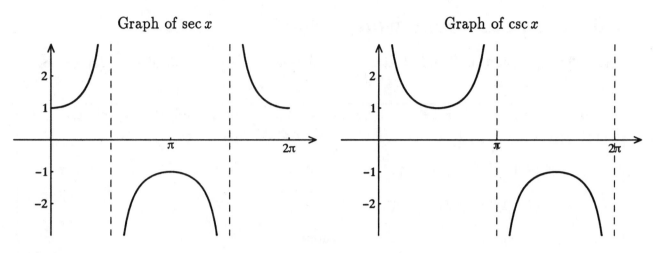

Remarks:

(R1) Domain: all real numbers that make sense (remember, no division by 0).

Range: tangent, cotangent: all reals

Range: secant, cosecant: $(-\infty, -1] \cup [1, \infty)$

(R2) The tangent, cotangent, secant, and cosecant all have vertical asymptotes.

(R3) Some properties:

$$|\sec x| \geq 1 \qquad |\csc x| \geq 1$$

$$\tan(-x) = -\tan(x): \quad \text{odd function}$$

$$\sec(-x) = \sec(x): \quad \text{even function} \hspace{4cm} \triangle$$

1.6   New functions from old

The following definition demonstrates how functions can be combined in order to obtain another function.

Definition: Let f and g be two functions with domains A and B, respectively. The functions $f + g$, $f - g$, $f \cdot g$, and f/g are defined as follows:

$$(f + g)(x) = f(x) + g(x) \qquad \text{domain } = A \cap B$$

$$(f - g)(x) = f(x) - g(x) \qquad \text{domain } = A \cap B$$

$$(f \cdot g)(x) = f(x) \cdot g(x) \qquad \text{domain } = A \cap B$$

$$\left(\frac{f}{g}\right)(x) = \frac{f(x)}{g(x)} \qquad\qquad \text{domain } = \{x \in A \cap B \mid g(x) \neq 0\}$$

Note!

(N1) The $+$ sign in $(f + g)(x)$ designates the operation of addition of functions.

The $+$ sign in $f(x) + g(x)$ denotes the usual addition of the real numbers $f(x) + g(x)$.

(Similarly for $(f - g)(x)$, $(f \cdot g)(x)$, and $(f/g)(x)$.)

(N2) The function f/g is not defined for those $x \in A \cap B$ such that $g(x) = 0$.   ◇

Example: Consider $f(x) = \dfrac{1}{\sqrt{x + 1}}$ and $g(x) = \sqrt{9 - x^2}$.

(E1) Find the domain of f.

(E2) Find the domain of g.

(E3) Find all the values of x common to the domains of f and g.

(E4) Find the function $f + g$.

(E5) Find the function $f - g$.

(E6) Find the function $f \cdot g$.

  Note! Display a graph of $f \cdot g$ on a calculator / computer.   ◇

(E7) Find the function f/g.

  Note! Display a graph of f/g on a calculator / computer.   ◇

□

Note! These algebraic operations also work on functions that do not have symbolic formulas. For example, f and/or g may be defined by a table, or graphically.   ◇

Definition: Let f and g be functions. The **composition** of f and g, denoted $f \circ g$, is the function defined by the rule
$$(f \circ g)(x) = f(g(x))$$

Remarks:

(R1) To obtain the value $(f \circ g)(x)$, first g operates on x to obtain the value $g(x)$ and then f operates on the value $g(x)$ to obtain the value $f(g(x))$.

(R2) The domain of $f \circ g$ are those x in the domain of g (hence, $g(x)$ is defined) such that $g(x)$ is in the domain of f (hence, $f(g(x))$ is defined).

(R3) Here is a diagram.

(R4) $\left\{ \begin{array}{ll} f \circ g : & \text{composition of } f \text{ and } g \\ g \circ f : & \text{composition of } g \text{ and } f \end{array} \right\}$: the *order is important* $\triangle$

Example: Consider $f(x) = \sqrt{x + 1}$ and $g(x) = x^2$.

(E1) Find the domain of f.

(E2) Find the domain of g.

(E3) Find the function $f \circ g$.

(E4) Find the function $g \circ f$.

□

Example: Consider the functions $f(x) = \sqrt{x+1}$, $g(x) = x^2$, and $h(x) = 1 - x$.

(E1) Find the function $f \circ h$.

(E2) Find the function $g \circ f \circ h$.

□

Note! Composition of functions by table, using graphs, or even by words is also possible. ◇

Definition: The **identity function** for composition is defined by the rule $I(x) = x$.

Note!

(N1) $f \circ I = I \circ f = f$

$$(f \circ I)(x) = f(I(x)) = f(x) \qquad \text{and} \qquad (I \circ f)(x) = I(f(x)) = f(x)$$

(N2) Composing any function f with I, in either order, gives f. ◇

Example: Consider the function h defined by $h(x) = \dfrac{3}{2 + \sqrt{x}}$. Find functions f and g such that $h(x) = (f \circ g)(x)$, where neither f nor g is the identity function.

□

Example: Consider the function h defined by $h(x) = \sin(x^2 + 3)$. Find functions f and g such that $h(x) = (f \circ g)(x)$, where neither f nor g is the identity function.

□

Two functions f and g are inverses if composing them, in either order, gives the identity function. The inverse of a function f is another function, denoted f^{-1}.

Example: Consider the functions $f(x) = 2x - 6$ and $f^{-1}(x) = \dfrac{1}{2}x + 3$.

(E1) Find the function $f^{-1} \circ f$.

(E2) Find the function $f \circ f^{-1}$.

$\square$

Definition: Let f and g be functions. If the equations

$$(f \circ g)(x) = x \qquad \text{and} \qquad (g \circ f)(x) = x$$

hold for all x in the domains of f and g respectively, then f and g are **inverse functions**. In this case, we write $g = f^{-1}$ (and $f = g^{-1}$).

Remarks:

(R1) If f and g are inverse functions, the domain of f is the range of g and the domain of g is the range of f.

(R2) If f and g are inverse functions, then

$$f(a) = b \quad \Longleftrightarrow \quad g(b) = a$$

for every a in the domain of f.

(R3) Here is a diagram.

(R4) Logarithm and exponential functions with the same base are the most important and useful examples of inverse functions. Recall, logarithms are defined as inverses of exponentials.

$$y = \ln x \iff x = e^y$$

(R5) If f and g are inverse functions, the graph of f is the reflection of the graph of g across the line $y = x$.

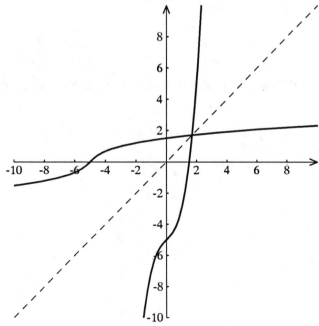

△

Question: What sorts of functions have inverses?

Definition: A function f is **one-to-one** if each y in the range of f is associated with a unique x in the domain of f.

Remarks:

(R1) If a function f is one-to-one, then we can determine a rule that assigns to each y in the range of f exactly one x in the domain of f.

(R2) If a function f is one-to-one, then we can determine an inverse function g.

(R3) A function f is one-to-one if and only if no horizontal line intersects the graph more than once.

   Here are a few examples.

(R4) Any function that is strictly monotonic is one-to-one, and hence, has an inverse.   ◇

Example: Let the function f be defined by $f(x) = \sqrt{x-1} + 2$.

(E1) Find the domain and the range of f.

(E2) Show f is one-to-one.

(E3) Find the function f^{-1}.

(E4) Here is a graph of f and f^{-1}. Which is which?

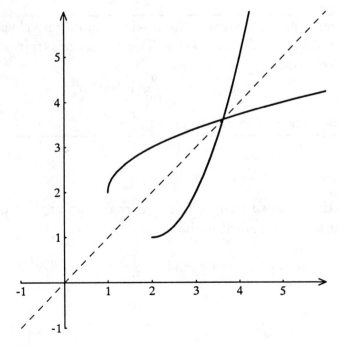

1.7   Modeling with elementary functions

Mathematical modeling: to model a phenomenon mathematically is to describe or represent the phenomenon quantitatively, in mathematical language. (Even elementary functions can be used to model real-world phenomena.)

Mathematical models have two important features.

(F1) Every model simplifies reality. We simply cannot describe every aspect of nature, or a machine, etc. A good model does describe general behavior or patterns with some accuracy.

(F2) A good model can be used for prediction.

Three steps to mathematical modeling.

(S1) Description: Describe the situation in mathematical language. Use functions, variables, equations, etc.

(S2) Deduction: See what logically follows from the mathematical description.

(S3) Interpretation: Interpret the mathematical results. Do they make sense in the context of the problem, what are the consequences, etc.

Fact: Suppose that a projectile leaves the origin with initial velocity v_0, aimed along the line $y = mx$, and influenced only by gravity. The projectile's trajectory is a parabola, given by the quadratic equation

$$y = mx - \frac{g}{2v_0^2}(1 + m^2)x^2.$$

(g is the (constant) acceleration due to gravity.)

Note!

(N1) If the projectile is aimed along a line that makes an angle θ with the positive x-axis, then $m = \tan \theta$ and the equation becomes

$$y = x \tan \theta - \frac{g}{2v_0^2}(1 + \tan^2 \theta)x^2$$

(N2) Use *consistent* units.

(U1) If distance is measured in feet, and v_0 in feet per second, then $g = 32$ f/s^2.

(U2) If distance is measured in meters, and v_0 in meters per second, then $g = 9.8$ m/s^2.

Example: Several years ago a highly touted but reclusive free agent Zen pitcher, Sid Finch, was clocked throwing a fastball with initial velocity 144 feet per second (roughly 98 miles per hour). The Red Sox coveted Sid, especially when they learned he could throw this heater along various lines $y = mx$. Consider some of the possible trajectories.

$$v_0 = 144, \quad g = 32$$

$$y = mx - \frac{1 + m^2}{1296} x^2 = -\frac{1 + m^2}{1296} x \left(x - \frac{1296m}{1 + m^2} \right)$$

Here are some trajectories for various values of m.

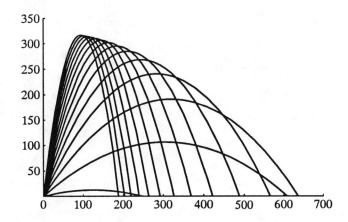

(E1) For what value of x does the ball hit the ground?

(E2) For what value of x does the ball reach it's peak?

(E3) What should the initial slope m be in order for Sid to throw the ball as high as possible? What initial slope gives the maximum possible *range*?

□

Example: A local organic farmer is concerned that the number of corn borers is increasing rapidly and destroying his crop. Over the last twenty years he has used a state approved procedure to count the number of pests per acre. In search of a pattern, he plots the data.

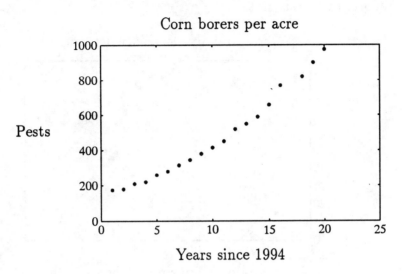

Our farmer has had calculus before, and tries to plot functions of the form $f(t) = ae^{bt}$. He eventually decides on the function $f(t) = 176e^{.0855t}$.

Here is a graph of the data along with the farmer's function.

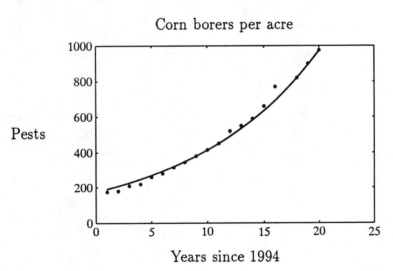

According to this graph, the next few years seem dismal. Can you predict the number of pests when $t = 25$?

□

Example: The demand for electricity in a small town naturally varies. The temperature, weather, and day of the week all affect the demand. However, averaged over many years, the daily demand looks familiar. Here is a plot of the average daily demand. The kilowatt hours (kwh) are reported in thousands.

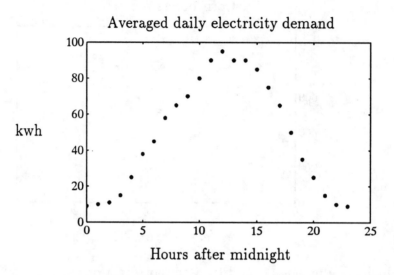

We might try to *fit* the data by a function that looks like $a\cos(bx + c) + d$.

Estimate the constants a, b, c, and d.

Here is a graph of the data together with $y = -42.5\cos\left(\dfrac{2\pi}{24}x\right) + 52.5$

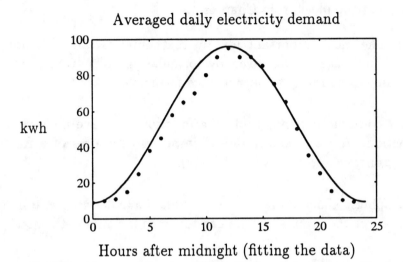

Averaged daily electricity demand

Hours after midnight (fitting the data)

Predict the electricity demand in this town at 6:00 am.

1.8   Chapter summary

Here is a summary of the ideas from Chapter 1.

(S1) Functions: The most important idea in mathematics (and Chapter 1) is that of a function. Functions may be given by formulas, graphs, tables, words, etc. In the real-world, functions rarely appear as nice formulas.

(S2) Graphs: By constructing the graph of a function, we have a geometric interpretation of the function. A graph shows lots of important features of a function (increasing, decreasing, concave up, concave down, etc).

(S3) Machine graphics: Calculators and computers make drawing graphs routine. We still need to interpret the graphs and be careful since machines still make mistakes.

(S4) Functions, formally: There are three parts to every function: rule, domain, and range.

(S5) Elementary functions: Algebraic, trigonometric, exponential, and logarithm functions and relatives are all important in calculus.

(S6) New functions from old: Sums, differences, products, quotients, composition, and inversion.

(S7) Modeling with elementary functions: Even though we call them elementary functions, they are still very useful in modeling real-world phenomena.

Chapter 2

The Derivative

2.1 Amount functions and rate functions: the idea of the derivative

Introduction

(I1) Calculus is all about change. The calculus tool that makes talking about change possible is the derivative. The basic idea is simple.

For a function f, the derivative function, denoted f', tells the rate of change of f.

(I2) In this chapter we take a look at the relationship between f and f'.
What does f' tell us about f?
How are the graphs of f and f' related?

(I3) Some functions may not have a nice derivative: it may not exist at certain points for example. In this chapter we will assume that f and f' are *nice*: f has a derivative f', and both f and f' are smooth, unbroken graphs.

The derivative as a rate function

(R1) Given any function f, we can always create or *derive* new functions based upon f.

(R2) Given all the possible relatives of f you can think of, the derivative function f' is the most important.

Definition: (The derivative as rate function.) Let f be any function. The new function f', called the **derivative** (or **rate function**) of f, is defined by the rule

$$f'(x) = \textbf{ instantaneous rate of change of } f \textbf{ at } x$$

Note!

(N1) *Derivative* and *rate function* will be used interchangeably to describe f'.

(N2) The function f is called an **original function** or an **amount function**.   ◇

Example: At 6:00 am on August 25, 1995 a car enters Interstate 95 heading north, from Boston. The car travels at varying speeds, sometimes north and sometimes south. Here are some definitions.

t : the elapsed time, measured in hours, since a reference time $t = 0$.

$P(t)$: the car's position at time t, measured in miles *north* of a reference point at time t.

$V(t)$: the cars *northward* velocity at time t, measured in miles per hour.

Here are possible graphs of the position and velocity functions.

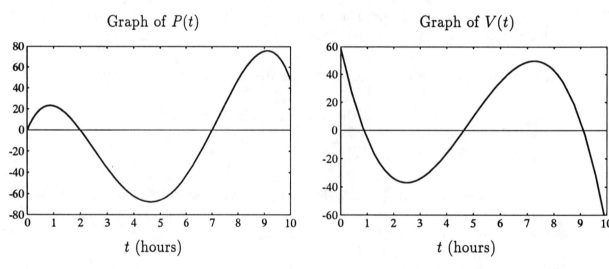

Remarks:

(R1) Reference point and reference time are arbitrary.

(R2) P and V are functions of t.

$P(8) =$

$V(8) =$

(R3) Think of P as an **amount function**, and V is the associated **rate function**. $V(t)$ returns the instantaneous rate of change of position with respect to time. $V(t) = P'(t)$.

(R4) Both P and V can be either positive or negative.

$P(3) =$

$V(3) =$

(R5) Position and velocity convey information about *direction*.

Distance and speed denote *nonnegative* quantities.

(R6) Speedometers can not tell direction.

So if the speedometer reads 49 at $t = 7$, then $V(7) = \pm49$.

But, the speedometer does measure **instantaneous speed**.

(R7) The rate of change of velocity with repsect to time is called **acceleration**. In symbols: $A(t) = V'(t)$. Acceleration can be positive or negative.

$A(5) > 0$: at $t = 5$ the car's northward velocity is increasing.

$A(8) < 0$: at $t = 8$ the car's northward velocity is decreasing. △

Note!

(N1) Just about any function f can be thought of as a rate or as an amount. Interpreting f' depends upon the context.

(I1) If $f(t)$ is the position at time t, then $f'(t)$ is the velocity at time t.

(I2) If $f(t)$ represents altitude, then $f'(t)$ is the rate of ascent.

(I3) If $f(t)$ denotes temperature, then $f'(t)$ is the rate of change of temperature at time t.

(N2) In general: $f'(x)$ represents the rate, at x, at which f grows with respect to x.

Suppose $f'(6) = 2.2$:
When $x = 6$ outputs from f increase 2.2 times as fast as inputs to f.

(N3) The proper units for measuring rates depend both on what quantities are varying and on the units used to measure those quantities.

Position: miles

Velocity:

Acceleration: ◇

The derivative graphically: f' as a slope function

(G1) What does the derivative mean graphically? How are the graphs of f and f' related?

(G2) Slope is the key. Slope is a rate of change.

The slope of the f-graph at any point $(x, f(x))$ is the instantaneous rate of change of f with respect to x.

Definition: (The derivative as a slope function.) Let f be any function. The **derivative** function f' is given by the rule

$$f'(x) = \text{the slope of the } f \text{ graph at } x$$

Example: Let $f(x) = 5x + 7$ (a linear function)

Then $f'(x) = 5$ Why?

Let's draw a graph of f and of f'.

Fact: For any linear function $f(x) = ax + b$, the rate function (or derivative) is the *constant* function $f'(x) = a$

The slope of a graph at a point: tangent lines

(T1) For straight lines, slope is easy: a line has only one slope.
      What about the slope of *any* graph at *any* point: this is a key idea in calculus!

(T2) The slope of a curve at a point P can be thought of as the slope of the tangent line at P. The tangent line *points in the direction of C..*

(T3) One can often draw reasonable tangent lines, and estimate their slopes.

Example: Here is the graph of a function f.

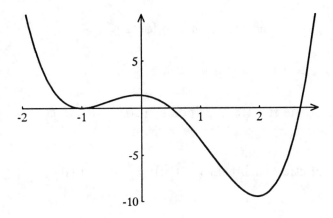

Draw some tangent lines and estimate the following.

$f'(1.5) =$

$f'(0) =$

$f'(1) =$

$f'(2) =$

$f'(2.5) =$ □

Relating amount and rate functions

(R1) How are f and f' related?

(R2) What can one function tell us about the other?

(R3) The rest of this section begins the discussion.

Fact: (The racetrack principle, or fast dogs win.) Let f and g be functions defined for all x in $[a, b]$, and suppose that $f(a) = g(a)$. If $f'(x) \leq g'(x)$ for all x in $[a, b]$ then $f(x) \leq g(x)$ for all x in $[a, b]$

Note!

(N1) $f(a) = g(a)$: f and g start together at $x = a$.

(N2) $f'(x) \leq g'(x)$: f grows no faster than g for $a \leq x \leq b$.

(N3) $f(x) \leq g(x)$: f never outruns g for $a \leq x \leq b$.

(N4) A *strict* version of the racetrack principle also holds: If $f(a) = g(a)$ and $f'(x) < g'(x)$ for x in $[a, b]$, then $f(x) < g(x)$ for every x in $(a, b]$. $\Diamond$

Example: Let f be a function such that (i) $f(0) = -5$; and (ii) $f'(x) \leq 3$ when $x \geq 0$. Show that $f(4) \leq 7$.

$\square$

Example: Let f be a function such that (i) $f(-2) = 10$; and (ii) $f'(x) \leq -1$ when $x \geq -2$. Show that $f(3) \leq 5$.

$\square$

Racetrack variants

(V1) If $f'(x) = g'(x)$ for $a \leq x \leq b$, then for some constant C, $f(x) - g(x) = C$ for $a \leq x \leq b$.

If f and g are growing at the same rate, then the distance between them must remain constant. If the derivatives of f and g are the same, then f and g differ by a constant.

(V2) If $f'(x) \leq g'(x)$ for $a \leq x \leq b$, then $f(x) - f(a) \leq g(x) - g(a)$ for $a \leq x \leq b$.

This variant deals with distance covered. If g is growing facter than f, then g must cover more distance over the same period of time.

(V3) If $f'(x) \leq g'(x)$ for $a \leq x \leq b$ and $f(b) = g(b)$ then $f(x) \geq g(x)$ for $a \leq x \leq b$.

g is growing faster than f, and just caught f at $x = b$. So f had to be ahead all of the time.

Here is a picture to illustrate this variant.

Example: Suppose $f(2) = 5$ and $f'(x) \leq 1$ for all x. What can be said about the values $f(4)$ and $f(-4)$.

$\square$

Example: Suppose that $f(0) = 2$ and $-3 \leq f'(x) \leq 4$ for all x. Find upper and lower bounds for $f(2)$. Find upper and lower bounds for $f(-2)$.

$\square$

Consider a falling object affected only by gravity (free fall).

> $h(t)$: the height of the object at any time t

> $v(t)$: the velocity of the object at any time t

> $a(t)$: the acceleration of the object at any time t

Facts:

(F1) $h'(t) = v(t)$; $v'(t) = a(t)$

(F2) $a(t) = v'(t) = g$ (Galileo)

> $g = -32 \text{ f/s}^2$ or $g = -9.8 \text{ m/s}^2$

(F3) Formulas for $h(t)$ (in feet) and $v(t)$ (in feet per second).

> $h(t) = h_0 + v_0 t - 16t^2$; $v(t) = v_0 - 32t$

Example: A ball is tossed from a bridge 40 feet high with an initial velocity of 55 ft/s

(E1) How high will the ball go?

(E2) When will the ball strike the ground?

(E3) What is the velocity of the ball at the instant it strikes the ground?

(E4) Graph h and v in reasonable viewing windows.

□

Note!

(N1) A **differential equation** or **DE** is an equation involving a function and some of its derivatives.

Examples:

$$f'(x) = -14$$

$$f'(x) = kf(x)$$

$$f''(x) + f(x) = e^x$$

(N2) As an object falls through the air there is really some air resistance. The deceleration due to air drag is related to the square of the velocity. Using derivative notation, this means

$$v'(t) = -g + kv(t)^2$$

where k is a positive constant that depends upon physical factors.

Taking air drag into account results in a more realistic model: as the object gathers speed, its acceleration tapers toward zero. $\diamond$

2.2   Estimating derivatives: a closer look

Recall: Given f and $x = a$.

(R1) $f'(a)$ is the instantaneous rate of change of f with respect to x at a.

(R2) $f'(a)$ is the slope of the f-graph at a.

In this section:

(S1) Take a closer look at f near a - *zoom in*.

(S2) Estimate numerical values of f'.

(S3) See more clearly what the derivative really means.

Estimating $f'(a)$: the slope of a graph at a point

(E1) Finding the slope of a line is easy. But what about the slope of a curve at a point?

(E2) The slope of a curve at a point P can be thought of as the slope of the curve's **tangent line at P**. This is the straight line through P that *points in the direction of the curve at P*.

Example: Let $f(x) = x^3 - 3x$. Use tangent lines to estimate $f'(-1.5)$ and $f'(.5)$. Interpret the results as rates of change.

Here is the graph of f, with reasonable tangent line at $x = 1$ and $x = -2$.

Graph of $f(x) = x^3 - 3x$ and two tangent lines

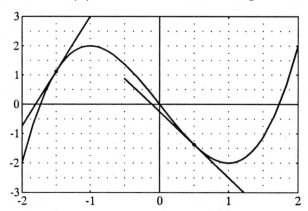

Estimate the slope of each tangent line from the graph.

What do the results mean in the language of rates of change?

□

Remarks:

(R1) Drawing a tangent line is one way to estimate the slope of a graph at a point.

(R2) We could also *zoom in* around the point in question until the graph *looks* straight.

(R3) This idea of zooming in almost always works. If it does, the function is called **locally linear** or **locally straight**. The graph of f looks like a line near $x = a$.   △

Example: Let $f(x) = x^3 - 3x$ as above. Estimate $f'(.5)$ by zooming in. Use the result to find an equation for the tangent line at $x = .5$.

Here is what happens if we zoom in on the f-graph near $(.5, f(.5)) = (.5, -1.375)$

Graph of $f(x) = x^3 - 3x$, close up   Graph of $f(x) = x^3 - 3x$, even closer

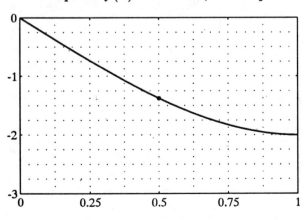
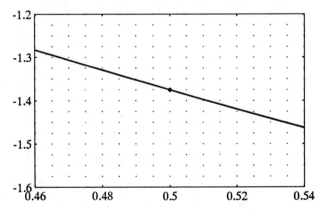

Estimate the slope of the tangent line, and find the equation of the tangent line at $x = .5$.

☐

Example: Let $f(x) = \cos x$. Is f locally linear at $x = .79$? Estimate the slope of the tangent line at $x = .79$.

Here are two graphs to help answer these questions.

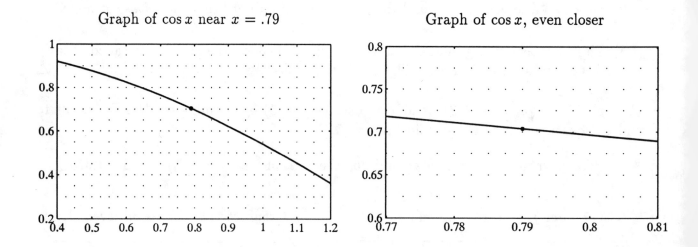

| Graph of $\cos x$ near $x = .79$ | Graph of $\cos x$, even closer |

☐

Example: Consider $f(x) = |x|$. See if you can find $f'(0)$ by zooming in. Does $f'(a)$ exist for $a \neq 0$?

Here are two graphs that might help.

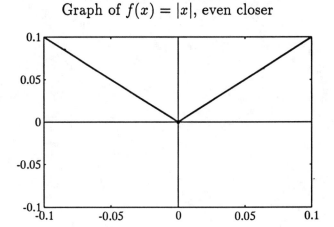

Graph of $f(x) = |x|$ Graph of $f(x) = |x|$, even closer

The graph of f has a sharp corner at $x = 0$. This *kink* or *corner* cannot be smoothed out by repeated zooming. So, it seems reasonable that the derivative does not exist at this kind of a point.   △ □

Plotting the rate function

(P1) Suppose we draw lots of tangent lines and estimate their slopes. So now we have lots of points of the form $(a, f'(a))$

(P2) If we connect the dots, we can get a rough sketch of the derivative function.

Example: For $f(x) = x^3 - 3x$, plot $f'(x)$. Make a guess at a formula for $f'(x)$.

Here is a table of values that might help, and a scatterplot.

x	-3	-2	-1.5	-1	0	$.5$	1	2	3
$f'(x)$	24	9	3.75	0	-3	-2.25	0	9	24

Graph of f'

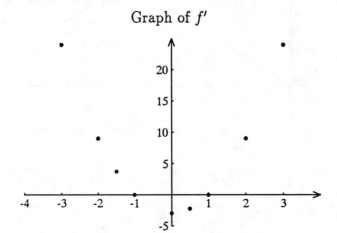

Example: Let $f(x) = e^x$. Use tangent lines to plot f' and guess a formula for f'. (Or, what's so *natural* about e^x?)

Complete this table of values by examining tangent lines

x	-2	-1	0	1	2
$f'(x)$					

Plot the points, connect the dots, and take a guess at a formula for f'

Graph of $f'(x)$

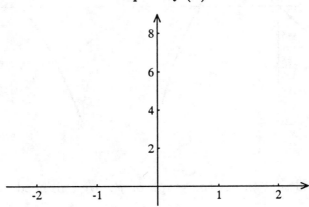

2.3   The geometry of derivatives

Recall: The geometry of a derivative is very easy to state: $f'(a)$ is the slope of the tangent line to the f-graph at $x = a$. (This is one of the most important statements in the course.)

Example: Consider the graphs of f and f' shown below.

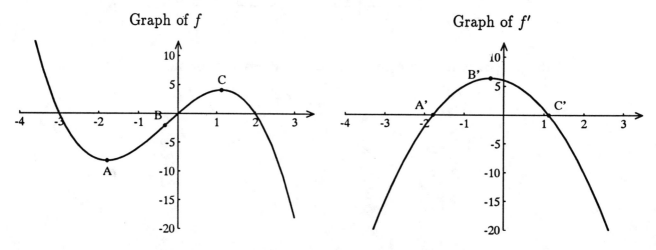

Remarks:

(R1) Have not proved that this is the graph of f'. We still need some analytic tools.

(R2) The graph of f' tells about the slopes of the tangent lines to the graph of f.

(R3) Consider some geometric relationships between the two graphs.   △

(E1) The sign of f'

    (S1) The graph of f falls to the left of A, rises between A and C, and falls again to the right of C.

    (S2) The sign of $f'(x)$ determines whether the line tangent to the graph of f at x points up or down.

    (S3) At A' and C' f' changes sign. So at A and C, f changes direction.

(E2) In the long run

    (L1) The graph of f' suggests that f increases without bound to the left of A and decreases without bound to the right of C.

    (L2) Geometrically: as $x \to \infty$ the graph of f falls more and more steeply.

(L3) Consider the graph of f' : to the right of C' f' is negative and decreasing.

(E3) Stationary, maximum, and minimum points

(M1) The points A and C are of particular interest: local minimum, local maximum (where the f-graph is horizontal).

(M2) $A = (-1.8, -8.2)$: $\qquad C = (1.1, 4)$

(M3) $x = 1.1$: local maximum point $\qquad f(1.1) = 4$: local maximum value of f

$\qquad x = -1.8$: local minimum point $\quad f(-1.8) = -8.2$: local minimum value of f

(M4) The x-coordinates of both A and C are called stationary points of f.

(E4) Any more stationary points?

(P1) There could be other stationary points on the graph of f.

(P2) We're not really sure if we have a complete graph yet.
(Here's where some analytic tools will help!)

(E6) Concavity and inflection

(C1) The point B, near $-1/3$, is an inflection point.

(C2) At B the concavity of f changes direction, from concave up to concave down.

(C3) Geometrically: at B the graph of f points most steeply upward. (Check this out by looking at the graph of f'.) $\qquad\qquad\qquad\qquad$ □

A function f **increases** when the graph rises, and **decreases** when the graph falls (from left to right). But what does all of this mean mathematically?

Definition: Let I denote the interval (a, b).

(D1) A function f is **increasing** on I if $f(x_1) < f(x_2)$ whenever $a < x_1 < x_2 < b$.

(D2) A function f is **decreasing** on I if $f(x_1) > f(x_2)$ whenever $a < x_1 < x_2 < b$.

Note!

(N1) If $f(x_1) \leq f(x_2)$ for $x_1 < x_2$, then f is **nondecreasing** on I.

(N2) If $f(x_1) \geq f(x_2)$ for $x_1 < x_2$, then f is **nonincreasing** on I.

(N3) In either case, f is **monotonic** on I.   ◇

Fact:

(F1) If $f'(x) > 0$ for all x in I, then f increases on I.

(F2) If $f'(x) < 0$ for all x in I, then f decreases on I.

Note!

(N1) This certainly seems reasonable and makes sense. Think about tangent lines, and where they are pointing, etc.

(N2) The proof of this *fact* requires the Mean Value Theorem.   ◇

Behavior at a point

(B1) We often say "f increases (or decreases) at the point $x = a$."

(B2) This really means that f increases (or decreases) is some (small) interval containing the point a.

(B3) So, here is the last fact restated.

Fact:

(F1) If $f'(a) > 0$, then f is increasing at $x = a$.

(F2) If $f'(a) < 0$, then f is decreasing at $x = a$.

Note!

(N1) The converse of this fact is *not* true!

(N2) We can*not* say: if f is increasing at $x = a$, then $f'(a) > 0$.   ◇

Example: Where is the function $f(x) = x + \sin x$ increasing? Where is $f'(x) > 0$? What happens at $x = \pi$? (Draw a graph and imagine a few tangent lines.)

□

So, we need to modify that fact again.

Fact:

(F1) If f increases at $x = a$, then $f'(a) \geq 0$.

(F2) If f decreases at $x = a$, then $f'(a) \leq 0$.

Here are a few more definitions involving local and global extreme values.

Definition: Let f be a function and x_0 a point of its domain.

(D1) x_0 is a **stationary point** of f if $f'(x_0) = 0$.

(D2) x_0 is a **local maximum point** of f if $f(x_0) \geq f(x)$ for all x is some open interval containing x_0. The number $f(x_0)$ is a **local maximum value** of f.

(D3) x_0 is a **global maximum point** of f if $f(x_0) \geq f(x)$ for all x in the domain of f. The number $f(x_0)$ is the **global maximum value** of f.

(Local and global **minimum points** are defined similarly.)

Note!

(N1) Points are inputs to a function. Values are the corresponding outputs.

(N2) Minimum (maximum) points may be local, global, or both.

(N3) **Extremum** (plural **extrema**): either maximum or minimum.
   We might say: the points A and C are local extrema of f.   ◇

So, how do we find local maximum, minimum points.

Fact: On a smooth graph every local maximum or local minimum point x_0 is a stationary point, i.e., a root of f'

Note!

(N1) To find maximum and minimum values of f, we need to search for roots of f'.

(N2) Be careful: $f'(x_0) = 0$ does *not* imply a maximum or minimum point at x_0.
   A stationary point may be a flat spot.   ◇

Example: The graph of a function f' appears below. Three points of interest are shown. Where, if at all, does f have local maximum or local minimum points? Why?

Graph of f'

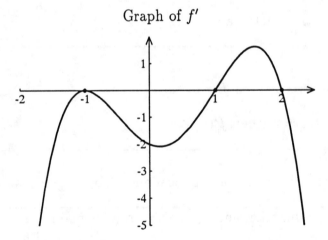

(E1) The three bullets on the graph, at $x = -1$, $x = 1$, and $x = 2$, represent roots of f'.

(E2) These points are stationary points of f.

(E3) Now the question is what kind of stationary point is each one: local max, local min, or flat spot?

(E4) We need to check the sign of f' around each stationary point.

(E5) A change in sign indicates a local extremum!! Why?

(P1) At $x = -1$

(P2) At $x = 1$

(P3) At $x = 2$

And here is the graph of f.

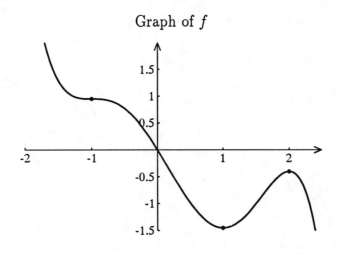

Graph of f

Example: The graph of a function f' appears below. Four points of interest are shown. Where, if at all, does f have local maximum or local minimum points? Why? Carefully sketch a possible graph of f.

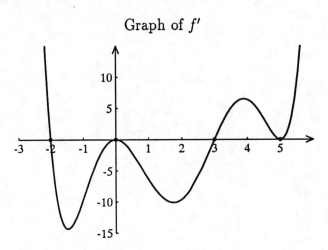

Graph of f'

Here's a summary of what we know about stationary points.

Fact: (First derivative test.) Suppose that $f'(x_0) = 0$.

(F1) If $f'(x) < 0$ for $x < x_0$ and $f'(x) > 0$ for $x > x_0$, then x_0 is a *local minimum point*.

(F2) If $f'(x) > 0$ for $x < x_0$ and $f'(x) < 0$ for $x > x_0$, then x_0 is a *local maximum point*.

There is still a case to settle: what happens when the graph of a function f has a *sharp edge* or *corner*?

Example: Consider the function $f(x) = |x - 1| + 2$. Sketch the graph of f and discuss any local (and global) extrema.

$\square$

Let's take a look at concavity more formally. Start by considering two graphs: one where f is concave up, one where f is concave down. Take a look at f' and try to describe how f' is changing.

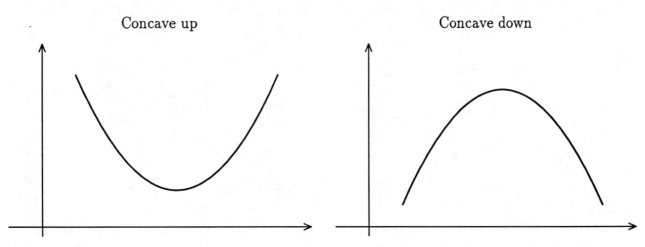

Concave up   Concave down

Definition:

(D1) The graph of f is **concave up** at $x = a$ if the slope function f' is increasing at $x = a$.

(D2) The graph of f is **concave down** at $x = a$ if f' is decreasing at $x = a$.

(D3) Any point at which a graph's direction of concavity changes is an **inflection point** of the graph.

Remark: An inflection point occurs wherever f' changes direction, wherever f' has a local maximum or local minimum.

Example: Here is a graph of f and a graph of f'. Discuss concavity and find all inflection points. Explain your results in derivative language.

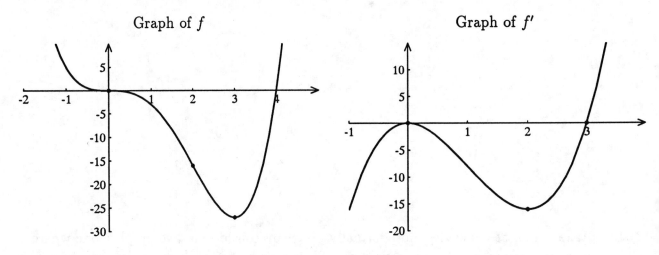

Graph of f Graph of f'

Example: The graph of the derivative of a function is shown below. Use the graph of f' to answer the following questions about f.

(E1) On which intervals is f increasing? Decreasing?

(E2) Where does f have stationary points?

(E3) Where does f have a local maximum? a local minimum?

(E4) On which intervals is f concave up? Concave down?

(E5) Where does f have a point of inflection?

(E6) Where does f achieve its maximum value on the interval $[0, 2]$? Its minimum value?

(E7) Where does f achieve its maximum value on the interval $[3, 6]$? Its minimum value?

(E8) Assume $f(0) = 0$. Sketch a graph of f.

Graph of f'

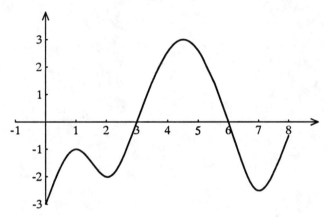

Example (continued):

□

2.4 The geometry of higher-order derivatives

Introduction

(I1) Recall: Given f : f' is the derivative of f.

(I2) f'' : **second derivative** of f, the derivative of f', also tells us some useful information about f.

Note!

(N1) The derivative of a function tells whether, and how fast, that function increases or decreases.

So, f'' tells whether, and how fast, the function f' increases or decreases.

(N2) If $f''(a) > 0$, then f' increases at a, and f is concave up at a.

If $f''(a) < 0$, then f' decreases at a, and f is concave down at a.   ◇

Fact: The graph of f is concave *up* where $f'' > 0$ and concave *down* where $f'' < 0$.

Note! If $f''(a) = 0$, no conclusion (concave up or concave down).   ◇

Example: Here is the graph of a function f (solid curve) and its second derivative f'' (dashed curve) on the same axes.

Graph of f and f''

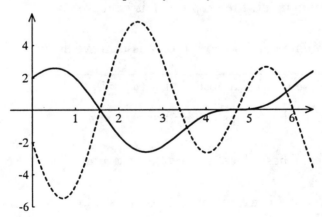

f is concave up precisely where $f'' > 0$, and concave down where $f'' < 0$.   □

© 1996 Harcourt Brace & Co.

Inflection points

(I1) f has an inflection point (IP) where the sign of f'' changes.

(I2) Every inflection point of f occurs at a root of f''.

(I3) $f''(a) = 0$ does *not* imply an inflection point at $x = a$.

> *Example* (This one is in the book also): Consider the graph of a function f (solid) and its derivatives f' (short dashes) and f'' (long dashes).

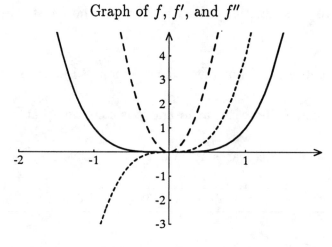

Graph of f, f', and f''

> $f''(0) = 0$, but $(0,0)$ is *not* an inflection point.
> The graph of f is concave up everywhere, f' is increasing everywhere.   □

Recall: a stationary point may be a local minimum, local maximum, or a flat spot. Concavity can help us decide.

(R1) At a local minimum point, the graph of f is concave up.

(R2) At a local maximum point, the graph of f is concave down.

Fact: (Second derivative test.) Suppose that $f'(a) = 0$.

(F1) If $f''(a) > 0$, then f has a local *minimum* at $x = a$.

(F2) If $f''(a) < 0$, then f has a local *maximum* at $x = a$.

(F3) If $f''(a) = 0$, anything can happen: f may have a local maximum, a local minimum, or neither at $x = a$, and the f-graph may be concave up, concave down, or have an inflection point.

Example: Consider $f(x) = -x^3 - x^2 + 6x$ (solid), $f'(x) = -3x^2 - 2x + 6$ (short dashes) and $f''(x) = -6x - 2$ (long dashes).

Graph of f, f', and f''

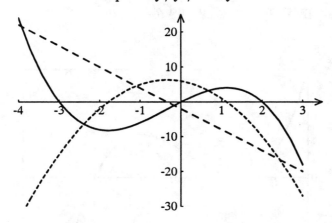

(E1) What does the derivative say about the f-graph?

(E2) How many stationary points are there? What are they?

(E3) Find any inflection points.

Remark: Don't try to memorize all of the cases. Just remember that f' tells whether, and how fast, the original function f increases or decreases.   △

Example: The graph of the *second* derivative of a function g is shown below. Use this graph to answer the following questions about g and g'.

Graph of g''

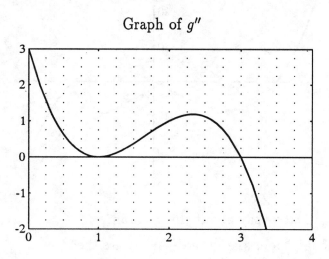

(E1) Where is g concave up?

(E2) Where does g have points of inflection?

(E3) Rank the four numbers $g'(0)$, $g'(1)$, $g'(2)$, and $g'(3)$ in increasing order.

(E4) Suppose that $g'(0) = 0$. Is g increasing or decreasing at $x = 2$? Justify your answer.

(E5) Suppose that $g'(0) = -4$. Is g increasing or decreasing at $x = 2$? Justify your answer.

(E6) Suppose that $g'(1) = -1$. Is g increasing or decreasing at $x = 2$? Justify your answer.

Example (continued):

$\square$

2.5   Average and instantaneous rates: defining the derivative

Introduction

(I1) We have defined f' informally. In this section we will look at the derivative more formally.

(I2) What is a tangent line exactly? What is instantaneous rate of change?

(I3) Given a function f, is there a general formula for f'?

Two problems

(P1) The rate of change problem: Traveling in a car with only an odometer, how can we find the instantaneous speed at any time t?

It seems reasonable to consider the average speed and use this to approximate instantaneous speed.

(P2) The tangent line problem: Given a curve C and a point P on C, describe the line tangent to C at P.

In a similar spirit, we might consider two points near P and approximate the tangent line with them.

Focus on the tangent line problem

(F1) Curves are often given as equations or as functions. Some curves are given only graphically or numerically. If the curve C has a formula, we want a formula for the tangent line to C at P.

(F2) Given C and P : all we need is the slope!

$$y = m(x - x_0) + y_0$$

(F3) We could estimate the slope. We can zoom in or draw a reasonable tangent line. (We've already done several example like this.)

(F4) Not every curve has a tangent line at every point P.

Consider $f(x) = |x|$. Remember that sharp corner at $x = 0$?

Two problems (more formally)

(P1) The tangent line problem: Let $y = f(x)$ be a function. Find the slope of the line
tangent to the graph of f at $x = 2$.

(P2) The instantaneous speed problem: Let the function $s = D(t)$ give the distance, in
miles, traveled by a car up to time t, in hours. Find the instantaneous speed at time
$t = 1$.

Remark: To solve these problems we need to understand the concept of a **limit**. $\triangle$

A solution to the tangent line problem

(T1) Given two points $P = (a, f(a))$ and $Q = (b, f(b))$ on a curve, we can draw a secant
line through P and Q. As P and Q get closer together (picture Q moving towards P),
the secant line gets closer to the tangent line. Here is a picture.

(T2) The slope of the tangent line is hard. But the slope of a secant line is easy. Given two
points on a line, find the slope, and use the point-slope formula for a line.

(T3) Slope of secant line $= m_{PQ} = \dfrac{\Delta f}{\Delta x} = \dfrac{f(b) - f(a)}{b - a}$

(T4) Slope of tangent line $= limit$ of slopes of secant lines over smaller and smaller intervals
$(m_{PQ} \to m)$.

Tangent line $=$ limiting position of the secant line (as $Q \to P$).

Limit: What is a function doing *near* a specific point. (more formal later)

Example: The point $P = (1, 5)$ lies on the graph of $f(x) = 3x^2 - 4x + 6$. Draw and defend a conclusion about $f'(1)$.

$x = 2$; $P = (1, 5)$, $Q = (2, 10)$

$m_{PQ} =$

$x = 1.5$; $P = (1, 5)$, $Q = (1.5, 6.75)$

$m_{PQ} =$

$x = 1.1$ $P = (1, 5)$, $Q = (1.1, 5.23)$

$m_{PQ} =$

x	2	1.5	1.1	1.01	1.001	0	.5	.9	.99	.999
m_{PQ}										

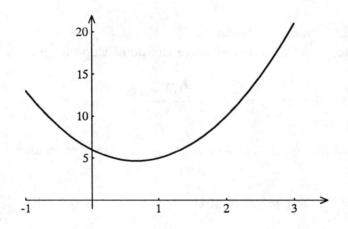

Example: Let $f(x) = 3x^2 - 4x + 6$. Consider the slope of the tangent line at $x = 1$ in symbols.

A solution to the instantaneous speed problem

(S1) Consider average speed:

$$\frac{\Delta D}{\Delta t} = \frac{D(1+h) - D(1)}{1 + h - 1} = \frac{\text{difference in position}}{\text{difference in time}}$$

(S2) The instantaneous speed at time $t = 1$ is the limiting value of the average velocities, over shorter and shorter time intervals, as $t = 1 + h$ moves closer to $t = 1$.

(S3) $D'(1) = \lim\limits_{h \to 0} \dfrac{D(1+h) - D(1)}{h}$

(S4) As $h \to 0$: average velocity (speed) $\to$ instantaneous velocity (speed)

Definition: Let f be a function defined near and at $x = a$. The **derivative of f at $x = a$**, denoted $f'(a)$ is defined by the limit

$$f'(a) = \lim_{h \to 0} \frac{f(a + h) - f(a)}{h}$$

Remarks:

(R1) $f'(a)$ may *not* exist.
   If $f'(a)$ does exist, f is **differentiable at $x = a$**, or f **has a derivative at $x = a$**.

(R2) Division by 0 :   must *do* something (algebra, some mathematical trick, etc.)

(R3) Other notation:

$$\lim_{h \to 0} \frac{f(a + h) - f(a)}{h} \; ; \qquad \lim_{\Delta x \to 0} \frac{f(a + \Delta x) - f(a)}{\Delta x} \; ; \qquad \lim_{x \to a} \frac{f(x) - f(a)}{x - a} \; ; \qquad \lim_{\Delta x \to 0} \frac{\Delta f}{\Delta x}$$

$\triangle$

Fact: For any function f, the ratio

$$\frac{f(a + h) - f(a)}{h}$$

is called a **difference quotient**. It measures the **average rate of change** of the function f from $x = a$ to $x = a + h$.

Remarks:

(R1) The denominator: $(a + h) - a = h = \Delta x$ represents a change in the input variable.

(R2) The numerator: $f(a + h) - f(a) = \Delta f$ represents a change in output.

(R3) The quotient is a ratio of two changes, a rate of change of f with respect to x.
   The sign indicates the direction.

(R4) The difference quotient is the *average* rate of change.

   The limit $\lim_{h \to 0} \dfrac{f(a + h) - f(a)}{h}$ is the *instantaneous* rate of change. $\triangle$

Difference quotients and derivatives: various settings

(D1) Slopes: Consider the graph of $y = f(x)$.

The difference quotient is the slope of the secant line.

The derivative $f'(a)$ is the slope of the tangent line at $x = a$.

(D2) Speeds: Let $y = f(x)$ be the position of an object at time x.

The difference quotient is the average speed.

The derivative $f'(a)$ is the instantaneous speed.

(D3) Waterworld: Let $y = f(x)$ be the total volume of water used by a household at any day x (since the beginning of the year).

The difference quotient represents the average rate of water used over an interval.

The derivative $f'(a)$ is the instantaneous rate of water usage.

2.6   Limits and continuity

Introduction

(I1) The notion of limits arose in the definition of the derivative, so now we will study limits more formally.

(I2) We will use the idea of limits to define and study continuity.

(I3) The basic idea of a limit is easy:

$\lim_{x \to a} f(x) = L$ means the limit of $f(x)$ as x approaches a, equals L.

$f(x)$ approaches L as x approaches a.

Remarks:

(R1) In symbols: $f(x) \to L$ as $x \to a$ for both $\left\{ \begin{array}{c} x < a \\ x > a \end{array} \right\}$ $x \neq a$

(R2) a, L : constants

(R3) f need *not* be defined at a. $\lim_{x \to a} f(x)$ examines the behavior of f near a.   △

Example: Let $f(x) = x + 1$. Find $\lim_{x \to 4} f(x)$.

□

Example: Let $f(x) = \dfrac{1}{1 + x^2}$. Find $\lim_{x \to 0} f(x)$.

□

Example: Let $f(x) = \dfrac{x^2 - 4}{x + 2}$. Find $\lim\limits_{x \to -2} f(x)$. Note that f is *not* defined at $x = -2$.

x	$f(x)$	x	$f(x)$
-3		-1	
-2.5		-1.5	
-2.1		-1.9	
-2.01		-1.99	
-2.001		-1.999	

If $x \neq -2$ then

$$f(x) = \frac{x^2 - 4}{x + 2} = \frac{(x - 2)(x + 2)}{x + 2} = x - 2$$

Find $\lim\limits_{x \to -2} x - 2$. Draw a graph to justify all of this.

□

Example: So, how does the derivative fit into all of this. Let $f(x) = x^2$. Use the limit definition to find $f'(-2)$.

□

Note! To solve a limit problem: use an algebraic trick, or a graphical solution, or a numerical tool.   ◇

Example: Let $f(x) = \tan x$. Find $f'(0)$.

Consider $f'(0)$ by definition.

Completing this table might help.

x	$\tan x / x$	x	$\tan x / x$
-1		1	
$-.5$		$.5$	
$-.1$		$.1$	
$-.01$		$.01$	
$-.001$		$.001$	

□

Example: (This one is in the book also, but it's a classic.) Let $f(x) = |x|$. Show using limits that $f'(0)$ does not exist.

□

Definition: (Limit, informally.) If $f(x)$ tends to a *single* number L as x approaches a, then $\lim_{x \to a} f(x) = L$.

Definition: (Limit, formally.) Suppose that for every positive number ϵ, no matter how small, there is a positive number δ so that

$$|f(x) - L| < \epsilon \quad \text{whenever} \quad 0 < |x - a| < \delta$$

Then

$$\lim_{x \to a} f(x) = L.$$

Remarks:

(R1) Consider $\quad 0 < |x - a| < \delta$

$\implies \quad -\delta < x - a < \delta$

$\implies \quad a - \delta < x < a + \delta \qquad x \in$ the open interval $(a - \delta, a + \delta)$

$0 < |x - a| \implies x \neq a, \quad f(a)$ need not be defined at a

(R2) Consider $|f(x) - L| < \epsilon$

$\implies$ $-\epsilon < f(x) - L < \epsilon$

$\implies$ $L - \epsilon < f(x) < L + \epsilon$ $f(x) \in$ the open interval $(L - \epsilon, L + \epsilon)$

(R3) In words: $f(x)$ can be arbitrarily close to L by taking x close enough to a.

(R4) Geometrically:

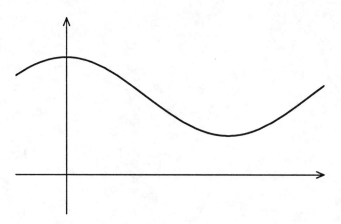

The graph of $y = f(x)$ lies between $L - \epsilon$ and $L + \epsilon$ when $a - \delta < x < a + \delta$

Any smaller nonzero δ will do.

(R5) The definition says L is the limit, but it does not say how to find it. $\triangle$

Note!

(N1) We can investigate limits numerically by evaluating $f(x)$ for x near a.

(N2) Again: $\lim\limits_{x \to a} f(x)$ may exist even if $f(a)$ is undefined.

(N3) If $f(a)$ is defined, no guarantee that $\lim\limits_{x \to a} f(x) = f(a)$ $\diamond$

Definition: (Right-hand limit.) $\lim\limits_{x \to a^+} f(x) = L$ means that $f(x) \to L$ as $x \to a$ from the *right.*

Definition: (Left-hand limit.) $\lim\limits_{x \to a^-} f(x) = L$ means that $f(x) \to L$ as $x \to a$ from the *left.*

Example: Let $f(x) = \begin{cases} x^2 - 1 & \text{if } x \leq 0 \\ 3 & \text{if } 0 < x < 2 \\ x + 1 & \text{if } x \geq 2 \end{cases}$

(E1) Carefully sketch the graph of $y = f(x)$.

(E2) Find $\quad \lim\limits_{x \to 2^-} f(x), \quad \lim\limits_{x \to 2^+} f(x), \quad \lim\limits_{x \to 2} f(x)$

(E3) Find $\quad \lim\limits_{x \to 0^-} f(x), \quad \lim\limits_{x \to 0^+} f(x), \quad \lim\limits_{x \to 0} f(x)$

$\square$

Fact: The ordinary limit $\lim\limits_{x \to a} f(x)$ exists if and only if both corresponding one-sided limits exist, and are equal. In symbols:

$$\lim_{x \to a} f(x) = L \iff \lim_{x \to a^+} f(x) = L = \lim_{x \to a^-} f(x).$$

© 1996 Harcourt Brace & Co.

Continuity

(C1) Roughly speaking, f is continuous if you can draw the graph without lifting your pencil. The graph has no break or jump.

(C2) Each of the following functions is continuous.

$$f(x) = x^3 + 3x + 5; \qquad g(x) = \cos x; \qquad h(x) = |x - 2|$$

The graph of h has a kink, but h is still continuous

Example: A function f is shown graphically below. Where is f continuous?

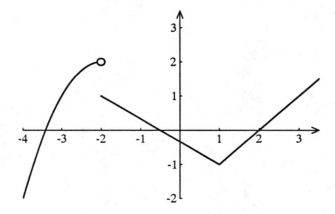

Definition: Let f be a function defined on an interval I containing $x = a$. If

$$\lim_{x \to a} f(x) = f(a)$$

then f is **continuous at** $x = a$. (If a is the left or right endpoint of I, then a right-hand limit or a left-hand limit replaces the two-sided limit above.) To say that f is continuous on I means that f is continuous at each point of I (including endpoints of I, if any).

Remarks:

(R1) Generally, elementary functions are continuous on their domains.

(R2) If f is continuous at $x = a$, then as $x \to a$, $f(x) \to f(a)$ △

Note! The definition for continuity is often broken into three parts. The function f is continuous at $x = a$ if:

(1) $f(a)$ is defined

(2) $\lim\limits_{x \to a} f(x)$ exists

(3) $\lim\limits_{x \to a} f(x) = f(a)$

Otherwise, f is **discontinuous** at $x = a$.   ◇

Example: A function f is shown graphically below. Consider continuity of f at $x = a, b, c$.

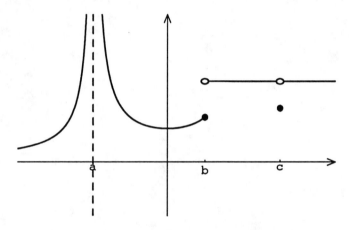

(E1) $x = a$: f is discontinuous at $x = a$, $f(a)$ is not defined.

  f has an **infinite discontinuity** at $x = a$. A one-sided limit is *not* finite.

(E2) $x = b$: f is discontinuous at $x = b$, $\lim\limits_{x \to b} f(x)$ does not exist.

  The two one-sided limits are different.

  f has a **jump discontinuity** at $x = b$. Both one-sided limits exist, but are unequal.

(E3) $x = c$: f is discontinuous at $x = c$, $\lim\limits_{x \to c} f(x) \neq f(c)$.

  The limit exists, but it is not equal to $f(c)$.

  f has a **removable discontinuity** at $x = c$.   □

Example: Let $f(x) = \dfrac{x^2 + x - 6}{x^2 - 4}$.

(E1) Show f is discontinuous at $x = 2$ and identify the discontinuity.

(E2) Show f is discontinuous at $x = -2$ and identify the discontinuity.

□

Example: Let $f(x) = \begin{cases} x^2 - 1 & \text{if } x \leq b \\ x & \text{if } x > b \end{cases}$

Find all values of b so that $\lim_{x \to b} f(x)$ exists.

□

2.7　Limits involving infinity; new limits from old

Introduction

(I1)　This section introduces another variation of limits, involving the symbol ∞.

(I2)　Limits at infinity: inputs increase/decrease without bound.

(I3)　Infinite limits: outputs increase/decrease without bound.

Definition: (Limit at infinity.) $\lim_{x \to \infty} f(x) = L$ means that $f(x) \to L$ as $x \to \infty$ (i.e., as x increases without bound).

Note!

(N1)　As x increases without bound, $f(x) \to L$.

(N2)　Values of $f(x)$ can be made arbitrarily close to L by taking x sufficiently large.

(N3)　L : a constant;　　∞ : *not* a number, indicates direction.

(N4)　Variant: $\lim_{x \to -\infty} f(x) = L$　　　　　　　　　　　　　　　　　$\Diamond$

Definition: (Infinite limit.) $\lim_{x \to a} f(x) = \infty$ means that $f(x) \to \infty$ (i.e., blows up) as $x \to a$.

Note!

(N1)　$f(x)$ increases without bound as $x \to a$.

(N2)　The limit does *not* exist, the symbol ∞ indicates the behavior of the function.

(N3)　Variants: use of $-\infty$, one-sided limits.　　　　　　　　　　　　　　$\Diamond$

Example: Consider the following limit statements. You might even try graphing these functions to help determine the limits.

(E1)　$\lim_{x \to \infty} \dfrac{2x - 1}{x} =$　　　　　　　(E2)　$\lim_{x \to \infty} \dfrac{\cos x}{x} =$

(E3)　$\lim_{x \to -\infty} \dfrac{x + 1}{x} =$　　　　　　　(E4)　$\lim_{x \to 2} \dfrac{1 - x^2}{(2 - x)^2} =$

(E5)　$\lim_{x \to 1+} \dfrac{x^2}{x^2 - 1} =$　　　　　　　(E6)　$\lim_{x \to 1-} \dfrac{x^2}{x^2 - 1} =$

$\square$

Definitions:

(D1) The line $y = L$ is a **horizontal asymptote** to the graph of f if either

$$\lim_{x \to \infty} f(x) = L \quad \text{or} \quad \lim_{x \to -\infty} f(x) = L$$

(D2) The line $x = a$ is a **vertical asymptote** to the graph of f if at least one of the following is true:

$$\lim_{x \to a} f(x) = \infty \qquad \lim_{x \to a^+} f(x) = \infty \qquad \lim_{x \to a^-} f(x) = \infty$$

$$\lim_{x \to a} f(x) = -\infty \qquad \lim_{x \to a^+} f(x) = -\infty \qquad \lim_{x \to a^-} f(x) = -\infty$$

Remarks:

(R1) Horizontal asymptotes reflect *long-run* behavior.
Vertical asymptotes reveal sudden spikes.

(R2) So, what kinds of functions have asymptotes anyway? $\triangle$

Fact: If $p(x)$ is a nonconstant polynomial, then $\lim\limits_{x \to \pm\infty} |p(x)| = \infty$.

Example:

(E1) $\lim\limits_{x \to \infty} (3x^3 + 2x^2 - 5x) =$

(E2) $\lim\limits_{x \to -\infty} (7x^5 - 2x^3 + 15) =$

$\square$

A word of caution: ∞ is *not* a number. So, arithmetic involving this symbol is tricky. Here are some typical expressions we will see and a few comments.

$1/\infty = 0$ Usually OK. This really means $\displaystyle\lim_{x\to\infty} \frac{1}{x} = 0$

$2 \cdot \infty = \infty$ OK. If a quantity increases without bound, so does $2 \cdot$ the quantity.

$\infty + \infty = \infty$ OK. If two quantities increase without bound, then so does their sum.

$1/0 \neq \infty$ Anything divided by zero is undefined.

$\infty - \infty \neq 0$ One quantity could dominate.

$\infty/\infty \neq 1$ In the context of a limit problem, it might be 1. But it could be some other number.

Note! To find horizontal asymptotes (limits at infinity) of rational functions, factor out the highest power of x from numerator and denominator. $\diamond$

Example: Find all horizontal asymptotes of the following functions:

(E1) $f(x) = \dfrac{1 - x^2}{1 + x^2}$

(E2) $g(x) = \dfrac{3x^2 - x - 2}{5x^2 + 4x + 1}$

(E3) $h(x) = \dfrac{x^2 - 2x}{x - 1}$

□

Note! Candidates for vertical asymptotes of rational functions are the roots of the denominator.   ◇

Example: Find all vertical asymptotes of the following functions. Try graphing each function.

(E1) $f(x) = \dfrac{2}{x - 3}$

(E2) $g(x) = \dfrac{x}{x^2 + x - 2}$

□

Example: Other algebraic functions may also have asymptotes. Find all the horizontal asymptotes of the following functions.

(E1) $f(x) = \dfrac{\sqrt{2x^2 + 1}}{3x - 5}$

(E2) $g(x) = x - \sqrt{x^2 + 1}$

□

In solving these limit problems, there are lots of background rules. Next up is a parade of theorems and facts.

Theorem 1. (Algebra with limits.) Suppose that

$$\lim_{x \to a} f(x) = L \quad \text{and} \quad \lim_{x \to a} g(x) = M$$

where L and M are finite numbers. Let k be any constant. Then

(i) $\lim\limits_{x \to a} k f(x) = k \lim\limits_{x \to a} f(x) = kL$

(ii) $\lim\limits_{x \to a}[f(x) + g(x)] = \lim\limits_{x \to a} f(x) + \lim\limits_{x \to a} g(x) = L + M$

(iii) $\lim\limits_{x \to a}[f(x) \cdot g(x)] = \left(\lim\limits_{x \to a} f(x)\right) \cdot \left(\lim\limits_{x \to a} g(x)\right) = L \cdot M$

(iv) $\lim\limits_{x \to a} \dfrac{f(x)}{g(x)} = \dfrac{\lim\limits_{x \to a} f(x)}{\lim\limits_{x \to a} g(x)} = \dfrac{L}{M}$ (if $M \neq 0$)

Remarks:

(R1) These limit laws may be stated in words. For example, the limit of a sum is the sum of the limits.

(R2) Each rule is reasonable, but the proofs require ϵs and δs.

(R3) These rules also hold for one-sided limits and limits at infinity. They hold for infinite limits ($L, M = \pm\infty$) if appropriate care is taken.

(R4) There are a few obvious limits that will help us find new limits.

(a) $\lim\limits_{x \to 0} |x| = 0$

(b) $\lim\limits_{x \to a} x = a$ for any real number a

(c) $\lim\limits_{x \to \infty} x^n = \infty$ for any $n > 0$

(d) $\lim\limits_{x \to \infty} \dfrac{1}{x^n} = 0$ for any $n > 0$ $\triangle$

Example: Show how the limit laws are used in solving the following problem.

$$\lim_{x \to 1} \frac{x^2 - x - 2}{x^2 + x - 6}$$

□

There are similar rules for continuous functions. Given some continuous functions, we can put them together to make other new continuous function.

Theorem 2. (Algebra with continuous functions.) Suppose that both f and g are continuous at $x = a$. Then each of the functions

$$f + g, \quad f - g, \quad \text{and} \quad fg$$

is also continuous at $x = a$. If $g(a) \neq 0$, then f/g is continuous at $x = a$.

Example: Try proving one part of Theorem 2.

□

Remember, another way to build new functions from old is through composition. So, there ought to be a technique to compute the limit of a composite function and to decide if the new composite function is continuous.

Fact: (Limit of a composite function.) Let f and g be functions, such that $\lim_{x \to a} g(x) = b$ and f is continuous at $x = b$. Then

$$\lim_{x \to a} (f \circ g)(x) = \lim_{x \to a} f(g(x)) = f(b)$$

Theorem 3. Suppose that g is continuous at $x = a$ and f is continuous at $g(a)$. Then $f \circ g$ is continuous at $x = a$.

Example: Find each limit.

(E1) $\lim_{x \to 4} \sqrt[3]{2x^2 - x - 1}$

(E2) $\lim_{x \to \pi} \cos(\sin x)$

(E3) $\lim_{x \to 3} \exp(-(x - 3)^2)$

$\square$

Remarks: Sometimes algebraic techniques do not help in solving limit problems. For example, $\lim_{x \to 0} x^2 \sin^2 \dfrac{1}{x}$

(R1) We can't split this one and use the limit laws since $\lim_{x \to 0} \sin^2 \dfrac{1}{x}$ does not exist.

(R2) A graph might help, but an analytical approach would be nice. $\triangle$

Theorem 4. (The squeeze principle.) Suppose that $f(x) \leq g(x) \leq h(x)$ for all x near $x = a$. If $f(x) \rightarrow L$ and $h(x) \rightarrow L$ as $x \rightarrow a$, then $g(x) \rightarrow L$ as $x \rightarrow a$.

Here is a graph to support this principle.

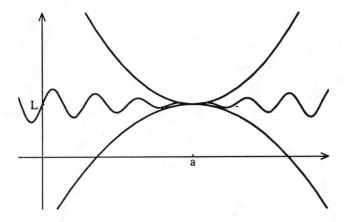

Note!

(N1) If the assumptions are met, the Theorem guarantees $\lim\limits_{x \to a} g(x)$ exists.

(N2) The hard part is finding the *bounding functions* f and h.   ◇

Example: Evaluate $\lim\limits_{x \to 0} x^2 \sin^2 \dfrac{1}{x}$.

□

Suppose we need to estimate $\lim_{x \to a} f(x)$

(E1) Consider a table of values near $x = a$.

(E2) Consider a graph of f near $x = a$.

Example: Estimate $\lim_{x \to 0} \left(\dfrac{1 - \cos x}{x} \right)$. Complete the table and sketch a graph.

x	$-.05$	$-.04$	$-.03$	$-.02$	$-.01$	$-.005$
$f(x)$						

x	$.05$	$.04$	$.03$	$.02$	$.01$	$.005$
$f(x)$						

2.8 Chapter summary

The derivative is the most important idea in calculus, and in Chapter 2.

(S1) Amount functions, rate functions, slope functions: Given an original function or amount function, the derivative may be interpreted as the rate function or as the slope function.

(S2) If $s(t)$ is the position of a moving object at time t, then $s'(t) = v(t)$ is the velocity, and $v'(t) = a(t)$ is the acceleration.

(S3) The racetrack principle: If g grows faster than f over an interval, then g grows more than f over that interval.

(S4) The slope of a curve at a point is the slope of the tangent line at that point. Zooming helps find the slope.

(S5) Derivatives, geometrically: The derivative f' of a function f tells whether and how fast the graph of f is rising.

(S6) The formal definition of derivative: $f'(x) = \lim_{h \to 0} \dfrac{f(x+h) - f(x)}{h}$

   (D1) Difference quotient: Slope of the secant line from x to $x + h$ or, the average rate of change of f from x to $x + h$.

   (D2) The derivative $f'(x)$: Slope of the tangent line at x or, the instantaneous rate of change of f at x.

(S7) Limits: $\lim_{x \to a} f(x) = L$ means $f(x)$ approaches L as x approaches a.

   (L1) A function is continuous if you can draw it without lifting your pencil. Continuity is defined in terms of limits.

   (L2) Infinite limits and limits at infinity involve the symbol ∞. These limits lead to horizontal and vertical asymptotes.

Example: The graph of the derivative of a certain function f is given below.

Graph of f'

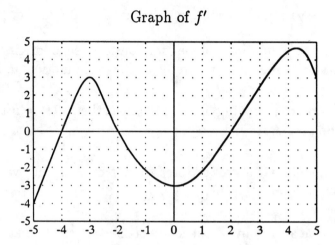

(E1) Suppose $f(-3) = 7$. Find an equation of the tangent line to the f-graph at the point $(-3, 7)$.

(E2) Estimate $f''(1)$.

(E3) At which values of x does the f-graph have inflection points?

(E4) On what interval(s) within $[-5, 5]$ is f increasing? decreasing?

(E5) Suppose $f(-4) = -8$. Find an upper bound for $f(-2)$.

□

Chapter 3

Derivatives of Elementary Functions

3.1   Derivatives of power functions and polynomials

Introduction

(I1) We have started the discussion of the derivative (rate function) and looked at the relationship between f and f'.

(I2) Geometrically: the secant line approaches the tangent line.

(I3) The average rate of speed approaches the instantaneous rate of speed.

(I4) Now, finally, it's time to learn a few analytical rules and calculate some derivatives. Given a function f, find f'. (You might think about going in the other direction: given f', find f.)

Notation: given $y = f(x) = x^3$.

(N1) The following notations all mean the same thing.

$$f'(x) = 3x^2 \, ; \qquad \frac{dy}{dx} = 3x^2 \, ; \qquad \frac{d}{dx}(x^3) = 3x^2 \, ; \qquad \frac{df}{dx} = 3x^2$$

(N2) These all mean the same thing too.

$$f'(2) = 12 \, ; \qquad \left.\frac{dy}{dx}\right|_{x=2} = 12 \, ; \qquad \left.\frac{d}{dx}(x^3)\right|_{x=2} = 12 \, ; \qquad \left.\frac{df}{dx}\right|_{x=2} = 12$$

Note! The $\dfrac{dy}{dx}$ form of the derivative is called **Leibniz notation**. It really signifies the definition of the derivative.   ◇

(N3) Here is some notation for higher order derivatives.

$$f''(x) = 6x\,;\qquad \frac{d^2y}{dx^2} = 6x\,;\qquad \frac{d^2}{dx^2}(x^3) = 6x\,;\qquad \frac{d^2f}{dx^2} = 6x$$

Definition: Let f be a function. The derivative of f, denoted f', is the function defined for an input x by

$$f'(x) = \lim_{h \to 0} \frac{f(x+h) - f(x)}{h}$$

Note!

(N1) The domain of f' is some *subset* of the domain of f. (The definition of f' requires f to be defined at x.) Usually, f and f' have the same domain, but sometimes the domain of f' omits a few points in the domain of f.

(N2) Finding the derivative by definition requires evaluating a limit. Usually you can not simply plug in $h = 0$. Rather you need some mathematical trick or clever algebraic manipulation.

(N3) If f is given by a graph or table, finding the limit may be impossible. We may still be able to get a numerical or graphical estimate.   ◇

Example: For each of the following functions, find the derivative by definition.

(E1) $l(x) = x$

(E2) $q(x) = x^2$

(E3) $c(x) = x^3$

(E4) $f(x) = \sqrt{x} = x^{1/2}$ Domain: $[0, \infty)$

(E5) $g(x) = \dfrac{1}{x} = x^{-1}$

$\square$

Remarks:

(R1) Many of these answers should seem familiar, we've drawn these graphs before. The actual limit calculations from definition now support our ideas.

(R2) So, what about a general formula for the derivative of x^k. If you study the above results, there seems to be a pattern here. $\triangle$

Theorem 1. (Power rule for derivatives.) Let k be any real constant. If $f(x) = x^k$, then $f'(x) = kx^{k-1}$.

Example: Find $f'(x)$.

(E1) $f(x) = x^{27}$

(E2) $f(x) = x^{\sqrt{2}}$

$\square$

Example: (What happens when you consider combinations of power functions?)
Let $p(x) = x^4 - 2x^3 - 7x^2 + x + 8$. Find $p'(x)$.

$\square$

Theorem 2. (The sum rule for derivatives.) Let f and g be differentiable functions and let $h = f + g$. Then

$$h'(x) = (f + g)'(x) = f'(x) + g'(x)$$

Theorem 3. (The constant rule for derivatives.) If k is any constant, f is any differentiable function, and $g(x) = kf(x)$, then

$$g'(x) = (k \cdot f)'(x) = k \cdot f'(x)$$

Note!

(N1) In words:

(W1) The derivative of a sum is the sum of the derivatives.

(W2) The derivative of a constant times a function is the constant times the derivative of the function. (Constants pass freely through derivative symbols.)

(N2) These theorems can be proved by going back to the definition of derivative (using limits).

(N3) Think geometrically. For example: $kf(x)$ is a vertical stretch. Everything gets multiplied by the same factor, k, even the tangent line.   ◇

Example: Find $f'(x)$.

(E1) $f(x) = x^3 - 2x + 4 + x^{1/3}$

(E2) $f(x) = 3x^5 - 17x^2 + 5x + 7$

(E3) $f(x) = (3x^2 + 1)^2$

□

Antiderivatives

(A1) Consider the opposite problem, **antidifferentiation**.
Given a function f, find another function F such that $F' = f$.
The function F is called an **antiderivative** of f.

(A2) In general, antidifferentiation is a harder problem than differentiation, except for polynomials and power functions. Those problems are pretty straight forward.

Example: Find an antiderivative for each of the following functions.

(E1) $f(x) = x + 2$

(E2) $g(x) = 3x^2 - 2x^{-2}$

(E3) $h(x) = 4x^3 - 3x^{-1/2} + x^{1/2}$

□

Note!

(N1) If F is an antiderivative for f, then so is $F + C$, for any constant C.

(N2) If $F'(x) = G'(x)$ for all x, then $F(x) = G(x) + C$ for some constant C.   ◇

Theorem 4. (Power rule for antiderivatives.) (Backwards power rule.) For any constants C and k ($k \neq -1$), $\dfrac{x^{k+1}}{k+1} + C$ is an antiderivative of x^k.

Note!

(N1) For a proof, just take the derivative of $\dfrac{x^{k+1}}{k+1} + C$

(N2) What about $k = -1$?

Fact: For $x > 0$, the natural logarithm function $\ln(x)$ is an antiderivative of $1/x$.

No proof yet, but take a look at the graphs of $1/x$ and $\ln x$.

Graph of $y = 1/x$ Graph of $y = \ln x$

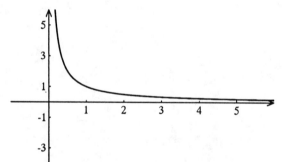
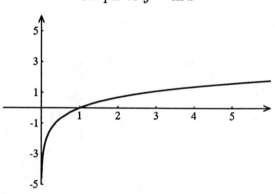

◇

Example: Find an antiderivative of $f(x) = \dfrac{1}{x} - 3x^5 + 6x^{3/4}$

□

Note!

(N1) It's easy to check if you have an antiderivative: just differentiate.

(N2) To find the antiderivative of a sum, take the antiderivative *term by term*. ◇

Example: The binomial theorem says that for any positive integer n

$$(a + b)^n = a^n + na^{n-1}b + \frac{n(n - 1)}{2}a^{n-2}b^2 + \cdots + nab^{n-1} + b^n$$

Use this theorem to show $\frac{d}{dx}(x^n) = nx^{n-1}$ for any positive integer n.

□

3.2   Using derivative and antiderivative formulas

Recall: For a falling object, let $h(t) =$ height of the object above the ground at any time t.

(R1) $v(t) = h'(t) =$ the velocity of the object

(R2) $a(t) = v'(t) = h''(t) =$ the acceleration of the object.

Suppose an object is in free fall, affected only by gravity.

(F1) $a(t) = -32 \text{ f/s}^2$

   v is an antiderivative of a, so $v(t) = -32t + C$

   $v(0) =$ initial velocity $= v_0 = -32(0) + C \implies C = v_0$

(F2) $v(t) = -32t + v_0$

   h is an antiderivative of v, so $h(t) = -32\dfrac{t^2}{2} + v_0 t + C = -16t^2 + v_0 t + C$

   $h(0) =$ initial height $= h_0 = C$

(F3) $h(t) = -16t^2 + v_0 t + h_0$

Example: A ball is thrown upward with speed 20 ft/s and strikes the ground 5 seconds later.

(E1) From what height was the ball thrown?

(E2) How high did the ball travel?

□

Example: A farmer has 1000 feet of fencing. He wants to enclose a rectangular area and subdivide it into four pens with fencing parallel to one side of the rectangle. What is the largest possible area of the four pens?

□

Example: Find the area of the largest rectangle that can be inscribed in the region above the x axis and bounded by the graph of $y = 4 - x^2$.

□

3.3   Derivatives of exponential and logarithm functions

Introduction

(I1) Using the definition of derivative to find f' for an algebraic function is pretty routine. Often we use an algebraic trick like multiplying by one in a convenient form, or finding a common denominator.

(I2) For non-algebraic functions, called **transcendental**, the usual algebraic tricks do not work. So, we need some other methods.

Recall: If $f(x) = e^x$ we saw some graphical evidence to suggest $f'(x) = e^x$. Does this work for every base b?

Given $g(x) = b^x$. $g'(x) = ?$ Here are two graphical approaches.

(G1) Plot several graphs: $y = 2^x$, $y = e^x$, $y = 3^x$, $y = 4^x$.

Moving left to right, all of the slopes start out small (and positive) and increase as x increases.

The graphs all cross at $x = 0$. But they all have different slopes at $x = 0$.

An interesting property seems to be that the slope of e^x at $x = 0$ is 1. Sometimes this property is used to define the base e.

Definition: The number e is the base for which the curve $y = b^x$ passes through $(0, 1)$ with slope 1.

(G2) For $g(x) = b^x$ where $b = 2$, estimate $g'(x)$ by looking at tangent lines, or use a calculator with a built-in procedure for estimating numerical derivatives. Plot $g'(x)/g(x)$. Try this for several values of b.

The ratio $g'(x)/g(x)$ turns out to be constant (related to b)!
This suggests that $g'(x) = kg(x)$.

So, what is this mysterious constant k?

Theorem 5. If $f(x) = e^x$, then $f'(x) = e^x$. If $g(x) = b^x$, then $g'(x) = b^x \ln b$.

Note!

(N1) The first formula is a special case ($b = e$) of the second.

(N2) The instantaneous rate of change of g is proportional to the value of g.   ◇

Take a look at $g'(x)$ by definition.

(D1) Why k exists. The key is to interpret the limit k as the slope of the line tangent to the g-graph at $x = 0$.

(D2) Why $k = \ln b$.

    We have already seen some pretty good numerical (and graphical evidence).
    Consider the case where $b = e$

Theorem 6. For any positive base $b \neq 1$, $(\log_b x)' = \dfrac{1}{x} \cdot \dfrac{1}{\ln b}$.

For the natural base $b = e$, $(\ln x)' = \dfrac{1}{x}$

In order to accept this theorem, remember that b^x and $\log_b x$ are inverses.

(A1) What b^x does, $\log_b x$ undoes.

(A2) We can obtain the graph of one by reflecting the other over the line $y = x$.

Here is a partial picture proof for Theorem 6.

(P1) The functions $f(x) = e^x$ and $g(x) = \ln x$ are inverses. Either graph can be obtained from the other by reflection through the line $y = x$. This graph illustrates how a reflection affects a tangent line.

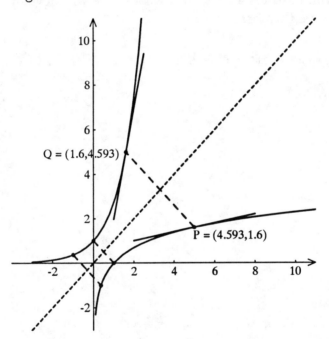

(P2) A point $P = (x, y)$ lies on the graph if and only if $Q = (y, x)$ lies on the other graph.

(P3) Consider tangent lines to P and Q. Like the graphs themselves, these tangent lines are symmetric with respect to the line $y = x$. Therefore: the slopes of the tangent lines at P and Q are reciprocals.

(P4) The slope of the tangent line to $f(x) = e^x$ at the point $Q = (\ln a, a)$: $e^{\ln a} = a$

Slope of tangent line to $g(x) = \ln x$ at the point $P = (a, \ln a)$: $1/a$

For every differentiation formula there is a corresponding antidifferentiation formula. Check out the following table.

$f(x)$	$F(x)$ (antiderivative)
e^x	$e^x + C$
b^x	$\dfrac{b^x}{\ln b} + C$
$\ln x$	$x \ln x - x + C$

Note!

(N1) The first two antiderivatives may be checked by taking the derivative.

(N2) The third formula seems a little mysterious. We will develop some calculus tools later on to help derive this formula.   ◇

Example: Let $f(x) = x^3 - 5x^2 + e^x - \ln x$.

(E1) Find $f'(x)$.

(E2) Find $f''(x)$.

(E3) Find an antiderivative of $f(x)$.

□

Example: Suppose A, and B are constants. Verify $y(x) = Ax^{-1} + B + \ln x$ is a solution to $x^2 y'' + 2xy' = 1$.

$\square$

3.4   Derivatives of trigonometric functions

Introduction

(I1) Graphical evidence suggests the following:

$$(\sin x)' = \cos x\,; \qquad (\cos x)' = -\sin x$$

(I2) To find the derivative of $\sin x$ we'll need two important limits:

(L1) $\displaystyle \lim_{h \to 0} \frac{\cos h - 1}{h} = \lim_{h \to 0} \frac{\cos h - \cos 0}{h - 0} = 0$

   Use some numerical and graphical justification for this. The second expression defines the derivative of $\cos x$ at 0, or the slope of the tangent line to $\cos x$ at 0.

(L2) $\displaystyle \lim_{h \to 0} \frac{\sin h}{h} = \lim_{h \to 0} \frac{\sin h - \sin 0}{h - 0} = 1$

   Use a similar argument here.

Derivative of $\sin x$:   Use the definition of derivative to find $(\sin x)'$.

□

Use a trigonometric identity to find $(\cos x)'$.

$\square$

Here is a summary table of trigonometric derivatives.

$f(x)$	$f'(x)$
$\sin x$	$\cos x$
$\cos x$	$-\sin x$
$\tan x$	$\sec^2 x$
$\csc x$	$-\csc x \cot x$
$\sec x$	$\sec x \tan x$
$\cot x$	$-\csc^2 x$

Note!

(N1) There are corresponding antiderivative formulas. For example

If $f(x) = \sec^2 x$ then $F(x) = \tan x + C$

(N2) Later, we will develop some rules for derivatives of products and quotients and *derive* the rest of these formulas. $\diamond$

Example: Let $f(x) = x^3 + 3x^{3/7} - \cos x + \tan x$. Find $f'(x)$.

$\square$

Example: Let $f(x) = 4x^7 + \sqrt{x} + \csc^2 x$. Find an antiderivative of f.

□

Example: Here is a graph of $f(x) = \sin x + \sqrt{3} \cos x$.

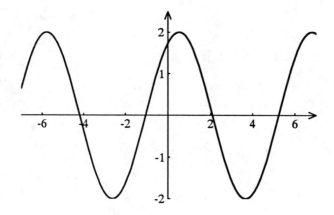

(E1) Find any local maxima or minima.

(E2) Find any inflection points.

□

Example: Let $f(x) = 2 - \sin x$ and $g(x) = 1 + \cos x$. Here is a graph of f and g. Which is which?

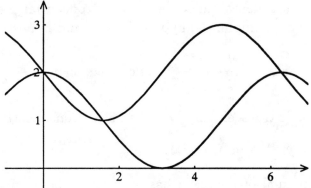

(E1) Find the maximum vertical distance between the curves $y = f(x)$ and $y = g(x)$ over the interval $[0, 2\pi]$.

(E2) Find the slopes of the tangent lines to f and g at the point where the maximum vertical distance occurs.

□

Some thoughts on transcendental functions

(T1) There are other transcendental functions besides exponential, logarithmic, and trigonometric functions. For example, the hyperbolic sine and hyperbolic cosine functions.

$$\sinh x = \frac{e^x - e^{-x}}{2} \quad \text{and} \quad \cosh x = \frac{e^x + e^{-x}}{2}$$

(T2) Showing a function is transcendental can be hard. Consider the sine function.

(S1) Could $\sin x = p(x)$, a polynomial?

No. A polynomial of degree n has at most n roots and the sine function has
infinitely many roots.

(S2) Could $\sin x = \dfrac{p(x)}{q(x)}$, a rational function?

No. A rational function has only finitely many roots.

(S3) $\sin x$ is periodic. There is no elementary function like that.

3.5 New derivatives from old: the product and quotient rules

Introduction

(I1) A common theme in calculus is producing new functions from old. Properties of the new functions (like continuity) are usually related to properties of the original, old functions.

(N2) Here we will take a look at derivatives of new functions: products and quotients. It seems reasonable that the derivative of the product (quotient) of two functions ought to be related to the derivative of the original functions.

Example: Find $f'(x)$.

(E1) $f(x) = x^5 + 2e^x + \ln x.$

$$f'(x) = 5x^4 + 2e^x + \frac{1}{x}$$

(E2) $f(x) = \pi \cos x + 3x^{-4/5}.$

$$f'(x) = \pi(-\cos x) + 3 \cdot -\frac{4}{5} x^{-\frac{4}{5}-1}$$

$$f'(x) = \pi(-\cos x) + -\frac{12}{5} x^{-9/5}$$

(E3) $f(x) = 4\sqrt[3]{x^2} + \sec x.$

$\square$

The derivative of a sum is pretty easy. Recall: the derivative of a sum is the sum of the derivatives. Products and quotients aren't so straight-forward.

Theorem 9. (The product rule.) If $p(x) = u(x) \cdot v(x)$, then

$$p'(x) = (u(x) \cdot v(x))' = u'(x) \cdot v(x) + u(x) \cdot v'(x).$$

Equivalently, $(uv)' = u'v + uv'$.

Note!

(N1) The derivative of a product is *not* the product of the derivatives: $(uv)' \neq u'v'$

(N2) The product rule can be extended to a product of three functions:

$$(uvw)' = u'vw + uv'w + uvw'$$

$\diamond$

Theorem 10. (The quotient rule.) If $q(x) = u(x)/v(x)$, then

$$q'(x) = \left(\frac{u(x)}{v(x)}\right)' = \frac{v(x)u'(x) - u(x)v'(x)}{v(x)^2}.$$

Equivalently, $\left(\dfrac{u}{v}\right)' = \dfrac{vu' - uv'}{v^2}.$

Note! The derivative of a quotient is *not* the quotient of the derivatives: $\left(\dfrac{u}{v}\right)' \neq \dfrac{u'}{v'}$ ◇

Example: Show $F(x) = x\ln x - x + C$ is an antiderivative of $\ln x$.

$u'(x) = 1 \quad u(x) = x$

$v'(x) = \dfrac{1}{x} \quad v(x) = \ln x$

$F'(x) = u'v + uv' - x' + C$

$F'(x) = 1 \cdot \ln x + x \cdot \dfrac{1}{x} - 1$

$F'(x) = \ln x + x\dfrac{1}{x} - 1$

$F'(x) = \ln x$

□

Example: $f'(x) = ?$ if:

(E1) $f(x) = (x^2 + 3x)(\cos x)$ $u(x) = x^2 + 3x \quad v(x) = \cos x$

$u'(x) = 2x + 3 \quad v'(x) = -\sin x$

$f'(x) = u'v + uv'$

$f'(x) = (2x+3)(\cos x) + (x^2 + 3x)(-\sin x)$

(E2) $f(x) = \sin x \cdot e^x$

$$f'(x) = u'v + uv'$$

$$f'(x) = (\cos x)(e^x) + (\sin x)(e^x)$$

$u(x) = \sin x \qquad v(x) = e^x$
$u'(x) = \cos x \qquad v'(x) = e^x$

(E3) $f(x) = \dfrac{x^2 + 4x}{x^3 - 5}$

$u(x) = x^2 + 4x \qquad v(x) = x^3 - 5$
$u'(x) = 2x + 4 \qquad v'(x) = 3x^2$

$$f'(x) = \frac{vu' - uv'}{v^2}$$

$$f'(x) = \frac{(x^3 - 5)(2x + 4) - (x^2 + 4x)(3x^2)}{(x^3 - 5)^2}$$

(E4) $f(x) = \dfrac{\ln x}{\sin x} - x^2$

$u(x) = \ln x \qquad v(x) = \sin x$
$u'(x) = \dfrac{1}{x} \qquad v'(x) = \cos x$

$$f'(x) = \frac{vu' - uv'}{v^2} - (x^2)'$$

$$f'(x) = \frac{(\sin x)(\tfrac{1}{x}) - (\ln x)(\cos x)}{(\sin x)^2} - 2x$$

□

Example: We have already seen a table with the derivatives of all the trigonometric functions. Now you can verify each formula. Let $f(x) = \sec x$. Find $f'(x)$.

$$f'(x) = (\sec x)'$$

$$\sec x = \frac{1}{\cos x}$$

$$f'(x) = \left(\frac{1}{\cos x}\right)'$$

$$\frac{u}{v}$$

$$u(x) = 1 \qquad v(x) = \cos x$$

$$f'(x) = \frac{vu' - uv'}{v^2}$$

$$u'(x) = 0 \qquad v'(x) = -\sin x$$

$$f'(x) = \frac{(\cos x)(0) - 1(-\sin x)}{(\cos x)^2} = \frac{-1(-\sin x)}{(\cos x)^2} = \frac{\sin x}{(\cos x)(\cos x)}$$

$$= \sin x \sec x \qquad\qquad \square$$

Example: Find the equation of the tangent line to the graph of $y = (x^2 + 5x)(x^4 + x + 2)(1 - x)$ at $x = -1$.

$$\begin{array}{cc} 1 + -5 & 1 \neq 1 \\ -4 & 2 \quad 2 \\ & -16 \end{array}$$

$$u(x) = x^2 + 5x$$

$$f'(x) = uvw'$$

$$u'(x) = 2x + 5$$

$$f'(x) = u'vw + uv'w + uvw'$$

$$v(x) = x^4 + x + 2$$

$$f'(x) = (2x+5)(x^4+x+2)(1-x) + (x^2+5x)(4x^3+1)(1-x) + (x^2+5x)(x^4+x+2)($$

$$v'(x) = 4x^3 + 1$$

$$= (2(-1)+5)((-1)^4+(-1)+2))(1+1) + ((-1)^2+5(-1))(4(-1)^3+1)(1+1) + (-1)^2+5(-1))((-1)^4+(-1)+2)(-1)$$

$$w(x) = 1 - x$$

$$(3)(2)(2) + (-3)(-3)(2) + (-3)(2)(-1)$$

$$w'(x) = -1$$

$$= 12 + 18 + 6$$

$$y + 16 = 36(x - 1)$$

Example: Let $f(x) = \dfrac{e^x}{x^2+2}$. Find $f'(x)$.

$$\frac{u}{v}$$

$$f'(x) = \frac{vu' - uv'}{v^2}$$

$$u(x) = e^x \qquad v(x) = x^2 + 2$$
$$u'(x) = e^x \qquad v'(x) = 2x$$

$$f'(x) = \frac{(x^2+2)(e^x) - (e^x)(2x)}{(2x)^2}$$

□

Example: Let $g(x) = (x^2+4)^{-1}$. Find $g'(x)$.
(Look for a pattern here. This leads us into the next section.)

$$g'(x) = \frac{-1}{x^2+4}$$

$$u(x) = 1 \qquad v(x) = x^2 + 4$$
$$u'(x) = 0 \qquad v'(x) = 2x$$

$$g'(x) = \frac{(x^2+4)(0) - 1(2x)}{(2x)^2}$$

$$g'(x) = \frac{-2x}{4x^2} = -\frac{1}{2x}$$

□

3.6   New derivatives from old: the chain rule

Introduction

(I1) What about the derivative of the composition of functions?

For example: $f(x) = \cos(x^2)$; $g(x) = (x^2 + 3x)^{10}$; $h(x) = (\ln x + e^x)^{-3}$

(I2) How do we put f, g, f', and g' together in order to find $(f \circ g)'$?

The chain rule is the answer.

Theorem 11. (The chain rule.) Let f and g be well-behaved functions. Then

$$(f \circ g)'(a) = f'(g(a)) \cdot g'(a)$$

Remarks:

(R1) In words: the derivative of the composite of two functions is the product of the derivative of the *outer* function evaluated at the *inner* function with the derivative of the inner function.

(R2) The theorem guarantees that the derivative $(f \circ g)'(a)$ exists, provided both $f'(g(a))$ and $g'(a)$ exist.

(R3) The theorem holds for any reasonable input a, so we can think of it as a relation among functions.

Therefore, here is some other notation: $(f \circ g)'(x) = f'(g(x)) \cdot g'(x)$

(R4) To find a derivative using the chain rule: start at the outside and work your way inside.

(R5) In Leibniz notation: Let y be a nice function of u, and u a nice function of x. Then y is a (composite) function of x; the derivative of y with respect to x is

$$\frac{dy}{dx} = \frac{dy}{du} \cdot \frac{du}{dx} \qquad\qquad \triangle$$

Example: Let $f(x) = (x^3 - 7x)^{-4}$. Find $f'(x)$.

$$f(u) = u^3 - 7u \qquad g(x) = x^{-4}$$

$$\left((x^3 - 7x)^{-4}\right)' = (f \circ g)'(x) = f'(g(x)) \cdot g'(x) = 3(x^{-4})^2 - 7 \quad \cdot \quad -4x^{-5}$$

$$= 3x^{-8} - 7 \cdot -4x^{-5}$$

□

Example: Let $f(x) = \sec(x^2)$ and $g(x) = \sec^2 x$. Find $f'(x)$ and $g'(x)$.

$$f(u) = \sec u \qquad g(x) = x^2$$
$$f'(u) = \sec u \, \tan u \qquad g'(x) = 2x$$

$$f'(x) = f'(g(x)) \cdot g'(x)$$

$$= \sec x^2 \cdot \tan x^2 \cdot 2x$$

$$f(u) = \sec u \qquad g(x) = x^2$$
$$f'(u) = \sec u \tan u \qquad g'(x) = 2x$$

$$g'(x) = g'(f(x)) \cdot f'(x)$$

$$g'(x) = 2(\sec x) \cdot \sec x \tan x$$

□

Example: Let $f(x) = \cos(\pi/x)$. Find an equation of the tangent line to the graph of f at $x = 1$.

$$f'(x) = -\sin\left(\frac{\pi}{x}\right) \cdot \pi$$

$$f'(1) = 0$$

□

Example: Let $g(x) = \left(\dfrac{x+3}{2x-1}\right)^7$. Find $g'(x)$.

$u(x) = x+3$ $g(u) = \dfrac{x+3}{2x-1}$ $\dfrac{u}{v}$

$h'(x) = 1$

$f(x) = x^7$

$f'(x) = 7x^6$

$f'(g(x)) \cdot g'(x) = g'(x)$

$g'(u) = \dfrac{vu' - uv'}{v^2}$

$v(x) = 2x-1$

$7\left(\dfrac{x+3}{2x-1}\right)^6 \cdot \dfrac{-7}{4x^2 - 4x + 1}$

$v'(x) = 2$ $g'(u) = \dfrac{(2x-1)(1) - (x+3)(2)}{(2x-1)^2}$

$(-x-3)(2)$

$(2x-1)(2x-1)$

$4x^2 - 2x - 2x + 1$

$4x^2 - 4x + 1$

$g'(u) = \dfrac{2x-1 - 2x - 6}{4x^2 - 4x + 1}$

$= \dfrac{-7}{4x^2 - 4x + 1}$

Example: Let $h(x) = \cos(\tan^2 x)$. Find $h'(x)$.

$h(u) = \cos x$ $h(x) = \tan x$ $f(x) = x^2$

$h'(u) = -\sin x$ $g'(x) = \sec x$ $f'(x) = 2x$

$(h \circ f \circ g) = h'(f(x)) \cdot f'(x) \cdot f'(g(x)) \cdot g'(x)$

$(h \circ f \circ g) = (h' \circ (f \circ g)) \cdot (f' \circ g \cdot g')$

$-\sin(\tan^2 x) \cdot 2(\tan x) \cdot \sec x$

$x^{\frac{1}{2}-1}$

Example: Let $f(x) = (x^2 + 7x)^\pi \sin\sqrt{x}$. Find $f'(x)$.

$\overset{u}{}$ $\overset{\checkmark}{}$

$f(u) = x^2 + 7x$ $g(x) = x^\pi$ $h(x) = \sin x$ $p(x) = \sqrt{x}$

$f'(u) = 2x+7$ $g'(x) = \pi x^{\pi-1}$

$g'(f(x)) \cdot f'(x) = \pi(x^2+7x)^{\pi-1} \cdot \cos\sqrt{x} \cdot x^{\frac{1}{2}}$

$h'(p(x)) \cdot p'(x) = \cos\sqrt{x} \cdot x^{-\frac{1}{2}}$

$f'(x) = u'v + uv'$

$f'(x) = \left(\pi(x^2+7x)^{\pi-1}\right)\left(\sin\sqrt{x}\right) + \left(x^2+7x\right)\left(\cos\sqrt{x} \cdot x^{-\frac{1}{2}}\right)$

Example: Let $g(x) = \sqrt[5]{x + \sqrt{x}}$. Find $g'(x)$.

$$g(u) = x + \sqrt{x} \qquad h(x) = x^{1/5} \qquad$$
$$g'(u) = 1 + x^{-1/2} \qquad h'(x) = \frac{1}{5} x^{-4/5}$$

$$h'(g(x)) \cdot g'(x) = \frac{1}{5}(x + \sqrt{x})^{-4/5} \cdot 1 + x^{-1/2}$$

□

Example: Let $h(x) = \sec^4(\sin^3 x)$. Find $h'(x)$.

$$h(x) = \sec x \qquad g(x) = x^4 \qquad p(x) = \sin x \qquad k(x) = x^3$$
$$= \sec x \cdot \tan x \qquad = 4x^3 \qquad = \cos x \qquad = 3x^2$$

$$(g' \circ (h \circ k \circ p)) = 4\left[\sec(\sin^3 x)\right]^3 \cdot \left[\sec(\sin^3 x) \cdot \tan(\sin^3 x)\right] \cdot 3(\sin x)^2 \cdot \cos x$$

$$(h' \circ (k \circ p))$$
$$(k' \circ p \cdot p')$$

□

Example: Let $f(x) = \cos(\sin(\tan x))$. Find $f'(x)$

$$f(u) = \cos u \qquad g(u) = \sin u \qquad h(u) = \tan u$$
$$f'(u) = -\sin u \qquad g'(u) = \cos u \qquad h'(u) = \sec u$$

$$(f' \circ (g \circ h)) \cdot (g' \circ h \cdot h') = -\sin(\sin(\tan x)) \cdot \cos(\tan x) \cdot \sec x$$

□

Example: Let $y = (f(x))^n$. Find y'.

Note! The result is called the general power rule. ◇

$$f \qquad g(x) = x^n \qquad h(x) = f(x)$$
$$g'(x) = nx^{n-1} \qquad =$$
$$y$$
$$y' = n(f(x))^{n-1} \cdot f'(x)$$

□

The chain rule can be used to relate the (known) derivative of a function to the (unknown) derivative of its inverse. Suppose f and g are inverse functions.

$$(f \circ g)(x) = f(g(x)) = x$$

$$(f \circ g)'(x) = f'(g(x)) \cdot g'(x) = 1$$

Fact: Let f and g be inverse functions. Then

$$g'(x) = \frac{1}{f'(g(x))}$$

for all x for which the right side is defined.

Remarks:

(R1) We will use this fact later on to find the derivatives of inverse trigonometric functions.

(R2) If f is differentiable with nonzero derivative, then its inverse, g, is differentiable. $\triangle$

Example: Let f and g be inverse functions with f defined by $f(x) = \sqrt{x-1} + 2$. Find $g'(9)$.

$\square$

3.7   Implicit differentiation

Introduction

(I1) Most functions we consider define y explicitly as a function of x: $y = f(x)$. These functions are easy to graph, and we have lots of rules for finding y'.

(I2) But curves in the xy-plane may also be represented by a relation involving both variables x and y, and it may be difficult (or impossible) to solve for $y = f(x)$. The given relation may define y implicitly as a function of x, or several functions of x.

(I3) So, if y is defined implicitly, how do we find y'?

Example: Consider the relation $x^2 + 9y^2 = 16$. This relation implicitly defines y as two functions of x. Find these two functions.

$$x^2 + 9y^2 = 16$$
$$9y^2 = 16 - x^2$$
$$\sqrt{y^2} = \sqrt{\frac{16 - x^2}{9}}$$
$$y = \pm \sqrt{\frac{16 - x^2}{9}}$$

$\square$

Example: Here is the graph of a relation. How can the variable y be defined as three functions of x?

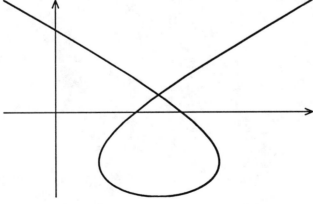

$\square$

Implicit differentiation: a procedure for finding $\dfrac{dy}{dx}$ if y is defined implicitly.

(S1) Treat y as a function of x (think of $y = f(x)$).

(S2) Differentiate both sides of the equation with respect to the variable x.

(S3) Solve for $\dfrac{dy}{dx}$, which may involve both x and y.

$$\frac{dy}{dx} = \lim_{\Delta x \to 0} \frac{\Delta y}{\Delta x}$$

Example: If $x^3 + y^2 + y^4 = 7$ find $\dfrac{dy}{dx}$.

$$3x^2 + y\frac{d \cdot y}{dy} + 3y^3 = 0$$

$$\frac{dy}{dx} = \frac{\left(-3y^5 - 3x^2\right)}{y}$$

$$3x^2 \frac{d}{dx} + y$$

$$\frac{d\,3x^2}{dx} + \frac{d\,y^2}{dx} + \frac{d\,y^4}{dx} = \frac{d\,7}{dx}$$

$$3x + y^2\frac{dy}{dx} \quad y^4\frac{dy}{dx} = 7$$

Example: If $x^3 y^2 = \cot(x - y)$ find $\dfrac{dy}{dx}$.

$$3x^2\,2y = \frac{dx}{dy} - \csc^2(x - y)$$

$$\frac{3x^2 2y}{-\csc^2(x - y) + y} = \frac{dx}{dy}$$

Example: Consider $x^3y^2 = x + y + 1$. At $(1, -1)$ find an equation of the tangent line, the normal line.

1. $m = y'(1) = ?$

2. $y(1) = ?$ $(1, y(1))$

$$3x^2 y^2 + x^3 2y \cdot y' = 1 + y' + 0$$
$$\underset{u}{\text{"}} \quad \underset{v}{\text{v}} + \underset{u}{\text{u}} \quad \underset{v'}{\text{v}'}$$

$$3x^2 y^2 - 1 = y'(1 - 2yx^3)$$

$$y' = \frac{3x^2 y^2 - 1}{1 - 2yx^3}$$

$$y'(1) = \frac{3y^2(1) - 1}{1 - 2y(1)} \qquad\qquad y'(-1) = \frac{3 - 1}{1 + 2} = \frac{2}{3}$$

$$y(1) = 2$$
$$y(1) = -1$$

$$\qquad\qquad\qquad\qquad y + 1 = \tfrac{2}{3}(x - 1)$$

□

3.8   Inverse trigonometric functions and their derivatives

Introduction

(I1) The problem is to define inverse trigonometric functions and find their derivatives.

(I2) The trigonometric functions are periodic and are not one-to-one functions on their natural domains.

(I3) To obtain one-to-one trigonometric functions we need to restrict the domains.

Inverse sine function

(S1) Let $f(x) = \sin x$ and restrict the domain: $[-\pi/2, \pi/2]$ (Range: $[-1, 1]$).

(S2) Restricted to this domain, the sine function is invertible.

Definition: For x in $[-1, 1]$, $\sin^{-1} x$ (or $\arcsin x$) is defined by the conditions

$$y = \sin^{-1} x \iff x = \sin y \quad \text{and} \quad -\frac{\pi}{2} \leq y \leq \frac{\pi}{2}$$

In words: $\arcsin x$ is the (unique) angle between $-\pi/2$ and $\pi/2$ whose sine is x.

Remarks:

(R1) Domain of $\sin^{-1} x$: $[-1, 1]$ Range of $\sin^{-1} x$: $[-\pi/2, \pi/2]$

(R2) $\sin^{-1}(\sin x) = x$ if $x \in [-\pi/2, \pi/2]$

$\sin(\sin^{-1} x) = x$ if $x \in [-1, 1]$

$\sin\left(\frac{\pi}{2}\right) = 1$

$\sin^{-1}(1) = \left(\frac{\pi}{2}\right)$

(R3) Here is a graph of the sine function and the arcsin function.

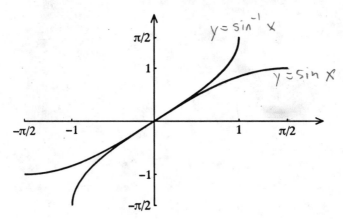

Example:

(E1) $\sin^{-1}\left(\dfrac{1}{2}\right) =$ $x = \sin\left(\frac{1}{2}\right)$ $\frac{1}{2} = \sin x,$ $x = \dfrac{\pi}{6}$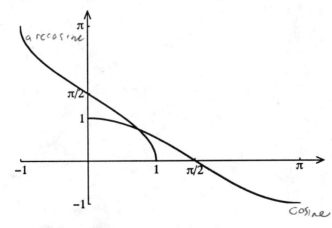

(E2) $\sin^{-1}\left(-\dfrac{1}{2}\right) =$ $-\dfrac{\pi}{6}$

(E3) $\sin\left(\sin^{-1}\left(\dfrac{3}{4}\right)\right) =$ $\sin\left(\sin^{-1}x\right) = x$ $= \dfrac{3}{4}$

(E4) $\sin^{-1}\left(\sin\left(\dfrac{7\pi}{4}\right)\right) =$ $\sin^{-1}\left(\sin x\right) = x$ $= \dfrac{7\pi}{4}$ □

Inverse cosine function

(C1) Let $f(x) = \cos x$ and restrict the domain: $[0, \pi]$ (Range: $[-1, 1]$).

(C2) Restricted to this domain, the cosine function is invertible.

Definition: For x in $[-1, 1]$, $\cos^{-1} x$ (or $\arccos x$) is defined by the conditions

$$y = \cos^{-1} x \iff x = \cos y \quad \text{and} \quad 0 \leq y \leq \pi$$

In words: $\arccos x$ is the (unique) angle between 0 and π whose cosine is x.

Remarks:

(R1) Domain of $\cos^{-1} x$: $[-1, 1]$ Range of $\cos^{-1} x$: $[0, \pi]$

(R2) $\cos^{-1}(\cos x) = x$ if $x \in [0, \pi]$

      $\cos(\cos^{-1} x) = x$ if $x \in [-1, 1]$

(R3) Here is a graph of the cosine function and the arccosine function.

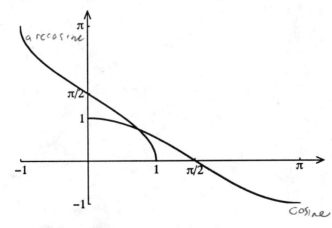

Example:

(E1) $\cos^{-1}\left(\dfrac{1}{2}\right) =$ 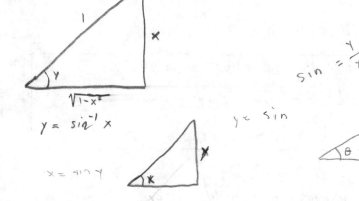 $\dfrac{60}{360}$ $\dfrac{2\pi}{6} = \dfrac{\pi}{3}$

(E2) $\cos^{-1}\left(-\dfrac{1}{2}\right) =$ $-\dfrac{\pi}{3}$

(E3) $\cos^{-1}\left(\cos\left(\dfrac{7\pi}{4}\right)\right) =$ $\cos^{-1}(\cos x) = x = \dfrac{7\pi}{4}$

(E4) $\cos(\sin^{-1}x) = $ $\dfrac{adj}{hyp} = \sqrt{1-x^2}$

$y = \sin^{-1}x$

$x = \sin y$

$\sin = \dfrac{y}{x}$

$y = \sin$

(E5) $\sin(\cos^{-1}x) = $ $\dfrac{opp}{hyp} = \dfrac{\sqrt{1-x^2}}{\sqrt{1-x^2}}$

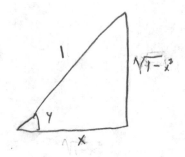

Inverse tangent function

(T1) Let $f(x) = \tan x$ and restrict the domain: $(-\pi/2, \pi/2)$ (Range: $(-\infty, \infty]$).

     Note! The restricted domain is an open interval. ◇

(T2) Restricted to this domain, the tangent function is invertible.

Definition: Let x be any real number. Then $y = \tan^{-1} x$ (or $y = \arctan x$) is defined by the conditions

$$y = \tan^{-1} x \iff x = \tan y \quad \text{and} \quad -\pi/2 < y < \pi/2$$

Remarks:

(R1) Domain of $\tan^{-1} x$: $(-\infty, \infty)$ Range of $\tan^{-1} x$: $(-\pi/2, \pi/2)$

(R2) $\tan^{-1}(\tan x) = x$ if $x \in (-\pi/2, \pi/2)$

     $\tan(\tan^{-1} x) = x$ if $x \in (-\infty, \infty)$

(R3) Here is a graph of the tangent function and the arctangent function.

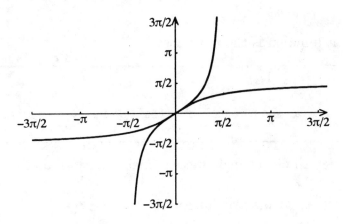

Example:

(E1) $\tan^{-1}(\sqrt{3}) =$ $\sqrt{3} = \tan x \quad = \quad 1.047$

(E2) $\tan^{-1}(-\sqrt{3}) =$ -1.047

(E3) $\tan^{-1}\left(\tan\left(\frac{7\pi}{4}\right)\right) =$ $\dfrac{7\pi}{4}$

(E4) $\cos(\tan^{-1} x) = \dfrac{adj}{hyp} = \dfrac{1}{\sqrt{1+x^2}}$

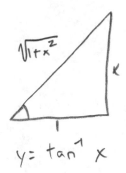

$y = \tan^{-1} x$

Remarks:

(R1) The remaining inverse trigonometric function definitions require care with domains and ranges.

(R2) The arcsecant function is the most useful.

> **Definition**: For any $|x| \geq 1$,
>
> $$y = \text{arcsec}\, x = \sec^{-1} x = \cos^{-1}(1/x) \iff x = \sec y$$

(R3) Remember that all of the angle measurements are in radians.   △

Derivatives of inverse trigonometric functions

$$y = \sin^{-1} x, \qquad y' = \frac{1}{\sqrt{1-x^2}} \qquad\qquad y = \cos^{-1} x, \qquad y' = \frac{-1}{\sqrt{1-x^2}}$$

$$y = \tan^{-1} x, \qquad y' = \frac{1}{1+x^2} \qquad\qquad y = \cot^{-1} x, \qquad y' = \frac{-1}{1+x^2}$$

$$y = \sec^{-1} x, \qquad y' = \frac{1}{|x|\sqrt{x^2-1}} \qquad\qquad y = \csc^{-1} x, \qquad y' = \frac{-1}{|x|\sqrt{x^2-1}}$$

Example: Let $f(x) = \cos^{-1} x$. From the table above, $f'(x) = \dfrac{-1}{\sqrt{1-x^2}}$.

(E1) Here is some graphical evidence to support this formula for $f'(x)$.

Graphs of $f(x) = \cos^{-1} x$ and $f'(x) = -1/\sqrt{1-x^2}$

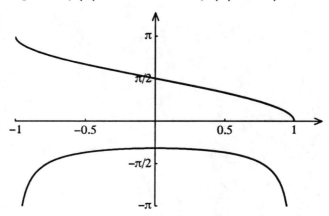

(E2) Prove $f'(x) = \dfrac{-1}{\sqrt{1-x^2}}$.

$$f(x) = \cos x \quad \text{and} \quad g(x) = \cos^{-1} x \qquad f'(x) = -\sin x$$

$$\left(\cos^{-1} x\right)' = \frac{1}{f'(g(x))}$$

$$= \frac{1}{-\sin\left(\cos^{-1} x\right)}$$

$$= -\frac{1}{\sqrt{1-x^2}}$$

$\square$

Example: Differentiate with respect to x.

(E1) $y = \sin^{-1}(x^3)$

$$f(x) = \sin^{-1} x = \frac{1}{\sqrt{1-x^2}}$$

$$g(x) = x^3 \qquad f'(g(x)) \cdot g'(x)$$

$$= \frac{1}{\sqrt{1-x^6}} \cdot 3x^2$$

(E2) $y = (\cos^{-1} 2x)^3$

$f(x) = \cos^{-1} x \qquad g(x) = x^3$

$\quad g'(f(x)) \cdot f'(x)$

$\qquad = 3\left(\cos^{-1} 2x\right)^3 \cdot \dfrac{-1}{2\sqrt{1-x^2}}$

(E3) $y = \sec^{-1}(2x^2 + 1)$

$y = \dfrac{1}{(2x^2+1)\sqrt{(2y^4 - 1)}}$

(E4) $x^2 y^3 + \tan^{-1}(x+y) = 3.$ Find $y' = \dfrac{dy}{dx}$

$\overset{\alpha}{\overset{\shortparallel}{}}$

$x + y = \tan \alpha$

$x^2 y^3$

$2xy^3 y' + x^2 3y^2 y' + \dfrac{1}{1+(x+y)^2} y' = 0$

3.9 Chapter summary

This chapter is all about derivatives and antiderivatives of elementary functions.

(S1) For a power function, the derivative is pretty easy. For a polynomial, take the derivative term by term.

(S2) Derivatives of exponential, logarithmic, and trigonometric functions are a little harder, but there are some nice formulas and patterns to follow.

(S3) For every differentiation formula, there is an antidifferentiation formula. In general, finding antiderivatives is harder than calculating derivatives. Finding antiderivatives is a major part of Calculus II.

(S4) The product rule and the quotient rule handle derivatives of products and quotients, and the chain rule is used for the derivative of the composition of functions.

(S5) As the name suggests, implicit differentiation is used to find the derivative of a function given implicitly.

(S6) With some special care for domains and ranges, we can define inverse trigonometric functions. Their derivatives look a little complicated but, surprisingly, are algebraic.

Chapter 4

Applications of the Derivative

4.1   Differential equations and their solutions

Basic ideas

(I1) A **differential equation** (**DE**) is any equation that involves a function and some of its derivatives.

Examples:

$$f'(x) = x^2 + 3x + 2\,; \qquad y' = ky\,; \qquad y'' = \sin x\,;$$

$$3xy - y^2 + 4xyy' = 0\,; \qquad \frac{d^2y}{dx^2} + e^{-x}\frac{dy}{dx} = 0$$

(I2) A **solution** to a DE is a function that satisfies the DE (a function that works). To **solve** a DE means to find a solution.

(I3) DE's are often stated without arguments. For example:

$$y' = ky\,, \qquad y'(x) = ky(x)\,, \qquad y'(t) = ky(t) \quad \text{(all the same DE)}$$

$$\frac{d^2y}{dx^2} - 4\frac{dy}{dx} + 13y = 2e^{2x}\cos 3x\,, \qquad y'' - 4y' + 13y = 2e^{2x}\cos 3x \quad \text{(the same DE)}$$

The input variable, or **argument** (x, t, etc.) is understood.

(I4) *Finding* a solution to a DE (acutally solving a DE) can be pretty tricky. *Checking* a solution is pretty straightforward: just see if y satisfies the equation.

Example: Solve the DE $f'(x) = x^2 + 3x + 2$. How many solutions are there? How are they related?

$$f(x) = \frac{1}{3}x^3 + 1.5x^2 + 2x + C$$

Infinite # of solutions, C is changing

□

Example: Solve the DE $y' = y$ (the function y is its own derivative). (This problem is in the text, but it's important enough to also do in class.) How many solutions are there? How are they related?

$$y = Ce^t$$

Infinate, C is changing

□

Initial value problems

(I1) **An initial value problem (IVP)** for a differential equation consists of finding a solution to the differential equation that also satisfies an **initial value condition** (usually of the form $y(x_0) = y_0$ or $f(x_0) = y_0$).

(I2) The initial condition can be used to find a particular value for the arbitrary constant obtained in solving the DE.

Example: Solve the IVP $f'(x) = x^2 + 3x + 2$; $f(0) = 17$.

$$f(x) = \tfrac{1}{3}x^3 + \tfrac{1}{2}x^2 + 2x + C$$

$$f(0) = \tfrac{1}{3}(0)^3 + \tfrac{1}{2}(0)^2 + 2(0) + C = 17$$

$$C = 17$$

$$\text{Solution} = \tfrac{1}{3}x^3 + \tfrac{1}{2}x^2 + 2x + 17$$

$\square$

Example: Solve the IVP $y' = y$; $y(0) = 17$.

$$y = Ce^t \qquad (e^x)' = e^x$$

$$y(0) = Ce^0 = 17 \Rightarrow C = 17$$

$$y = 17e^t \quad \text{is the unique solution to the IVP}$$

$\square$

Example: Show $y(t) = T + Ae^{kt}$ is a solution to the DE $y' = k(y - T)$, where k, T, and A are constants.

$$y' = k(y - T)$$

$$(T + Ae^{kt})' = k(y - T)$$

$$LHS = y'(t) = k \cdot Ae^{kt}$$

$$RHS = k(y - T) = k(T + Ae^{kt} - T) = k \cdot Ae^{kt}$$

$\square$

Newton's Law of Cooling

(N1) The rate at which an object cools is proportional to the temperature difference between the object and its environment, or

(N2) The relative rate of change in temperature difference between an object and its surrounding medium is a constant.

$y(t)$: temperature of the object at time t

T : temperature of surrounding medium; assumed to be a constant

$y' = k(y - T)$: and we just saw a solution to this DE in the last problem.

$y(t) = T + Ae^{kt}$ Ae^{kt} A

Example: An object with an initial temperature of 10° F warms to 16° F in one hour when placed in air with temperature of 32° F.

(E1) Find the temperature of the object after three hours.

$y(0) = 10$

$y(1) = 16$

$x(3) = 32 +$

$y' = k(y + 32)$

$10 = y(0) = 32 + Ae^{k \cdot 0}$

$y(0) = 32 + A$

$10 = A + 32$

$A = -22 \quad A(42)$

$y(3) = 32 - 22e^{-.31845(3)}$

$y(3) = 3.735$

$y(3) = 23.54°F$

$y(1) = 32 - 22e^{k(1)}$

$16 = 32 - 22e^{k}$

$-16 = -22e^{k}$

$e^{k} = \dfrac{16}{22} \quad \left(\dfrac{-26}{40}\right)$

$K = \dfrac{(\ln 16 - \ln 22)}{2}$

$y(t) = 32 + 40 e^{.21 539t}$

$k = .0835$

$y(t) = 32 - 22e^{-.31845}$

(E2) After how many hours was the object's temperature 28° F?

$$28 = 32 - 22e^{-.31845t}$$

$$-4 = -22e^{-.31845t}$$

$$\frac{-4}{-22} = e^{-.31845t}$$

$$\log_b(x^r) = r\log_b x$$

$$-1.705 = -.31845t \qquad -.31845$$

$$\log_e x = \ln x$$

$$t = 5.35 \text{ hours}$$

$$10^{3t} = 3$$

$$\log 3t$$

$$x$$

$$10^{2x} = 100$$

$$2x(\log 10) = \log 100$$

$$\frac{2x}{2} = \frac{2}{2}$$

$$x = 1$$

Here is a graph of the function $y(t) = 32 - 22e^{-.319t}$.

Example: Solve the DE $f'(x) = x^2 e^{x^3}$.

$u'v + uv'$

$x^2 \cdot e^{x^3}$ $f(x)$

$\frac{1}{3}x^3 \cdot e^{\frac{1}{3}x^4}$ e^{x^3}

$\frac{1}{3}x^3 + \frac{1}{3}x$

□

Example: Solve the DE $g'(x) = (\tan x)^3 \sec^2 x$.

□

Example: Solve the IVP $y' = .01y$; $y(0) = 35$.

$y = 100e^x - 65$

$y(0) = Ce^0$ $y' = .01(100)^x$

$35 = C$ $= e^x$

$y =$

□

Example: Solve the IVP $\dfrac{dy}{dx} = \dfrac{x^2}{1+y^2}$; $y(2) = 1$

$$y' = \frac{x^2}{1+y^2}$$

y

4.2   More differential equations: modeling growth

Exponential growth

(E1) Consider a mathematical model to describe the following:

The rate of change of y is proportional to y.

(E2) In calculus terms: $\dfrac{y'(t)}{y(t)} = k$ or $y'(t) = ky(t)$

k is sometimes called the constant of proportionality.

(E3) Exponential functions fit this model: they grow at a rate proportional to their size (the amount present).

Theorem 1. For any constants A and k, the exponential function $y = Ae^{kt}$ solves the initial value problem
$$y' = ky; \qquad y(0) = A$$

Example: A certain bacterial population is known to grow at a rate of six percent of its population each day. The initial size of the population is 5000 organisms.

(E1) How many organisms are present after 10 days?

$$y = Ae^{kt}$$
$$y = 5000e^{.06(10)}$$
$$y = 9110.59 \text{ organisms}$$

(E2) How long will it take for the initial population to double?

$$10000 = 5000e^{.06t}$$

$$2 = e^{.06t}$$

$$\ln 2 = .06t$$

$$t = \frac{\ln 2}{.06}$$

$$t = 11.55 \, days$$

□

Example: Upon the death of an organism, radioactive C^{14} disintegrates at a rate proportional to the amount present in the organism. The half-life of C^{14} is 5728 years.

(E1) Find the number of years until 30% of the amount of C^{14} has disintegrated.

$$y = Ae^{kt}$$

$$0.7 = 1e^{-.000121t}$$

$$\frac{\ln 0.7}{-.000121} = t$$

$$t = 2947.7 \, years$$

$$\frac{1}{2} = 1e^{k(5728)}$$

$$\ln \frac{1}{2} = 5728k$$

$$k = \frac{\ln \frac{1}{2}}{5728}$$

$$k = -.000121$$

(E2) Find the number of years until 20% of the amount of C^{14} remains.

$$0.2 = 1e^{-.000121t}$$

$$\frac{\ln 0.2}{-.000121} = t$$

$$t = 13301.04 \, years$$

□

Compound interest

(I1) When money is deposited in a savings account, **interest** is paid on the deposit at stated intervals.

(I2) The interest is added to the savings account, and the original deposit plus the interest both earn interest. This is called **compound interest**.

(I3) The original amount of money deposited is called the **principal amount**.

(I4) The principal amount plus the compound interest is the **compound amount**.

(I5) The time interval between interest payments is the **interest period**.

(I6) The **interest rate** i is expressed as an annual (yearly) rate in percent.

 For calculations, write i in decimal notation.

Notation

(N1) $b(t)$: the balance (in dollars) in the savings account at time t (in years).

(N2) $b(0)$: initial balance

Interest compounded annually

$$b(1) = b(0) + ib(0) = b(0)(1 + i)$$

$$b(2) = b(1) + ib(1) = b(1)(1 + i) = b(0)(1 + i)^2$$

$$b(3) = b(2) + ib(2) = b(2)(1 + i) = b(0)(1 + i)^3$$

$$b(t) = b(t - 1) + ib(t - 1) = b(t - 1)(1 + i) = b(0)(1 + i)^t$$

Example: If a bank pays 4% compounded annually, what principal is necessary in order to achieve a compound amount of $10000 in 10 years?

$$10000 = b(0)\left(1 + .04\right)^{10}$$

$$b(0) = \frac{10000}{(1 + .04)^{10}}$$

$$b(0) = \$6755.64$$

Interest compounded n times per year $(n \geq 1)$

Interest rate per period $= i/n$

The number of interest periods per year is n

Suppose interest is paid monthly $(n = 12)$. Here's what happens the first few months.

$$b\left(\frac{1}{12}\right) = b(0)\left(1 + \frac{i}{12}\right) : \text{ 1 month}$$

$$b\left(\frac{2}{12}\right) = b(0)\left(1 + \frac{i}{12}\right)^2 : \text{ 2 months}$$

So, $b(1) = b(0)\left(1 + \frac{i}{12}\right)^{12}$

And, $b(2) = b(0)\left(1 + \frac{i}{12}\right)^{24}$

In general: $b(t) = b(0)\left(1 + \frac{i}{n}\right)^{nt}$

Remarks:

(R1) $\left(1 + \frac{i}{n}\right)^n$: **effective annual interest rate**

(R2) $n = 52$: interest compounded weekly $\qquad n = 365$: interest compounded daily $\quad \triangle$

Interest compounded *continuously*

(C1) In reality, the balance $b(t)$ varies discretely, at regular intervals and in jumps of at least a penny.

(C2) However, if we try to model the (discrete) balance with a continuous function and think of interest as being paid at every instant, or continuously, then $b(t)$ must satisfy the DE $b'(t) = i \cdot b(t)$

(C3) The solution to this DE (Theorem 1) is $b(t) = b(0)e^{it}$

Example: How long does it take for an investment to double if the annual interest rate of 4.5% is compounded:

(E1) monthly? 15.43 years

(E2) weekly? 15.41 years

(E3) daily?   15.405 years

$$2 = 1\left(1 + \frac{.045}{12}\right)^{12t}$$

(E4) continuously? 15.403 years

$$\frac{\log 2}{\log\left(1 + \frac{.045}{12}\right)} = 12t$$

$$2 = (1.00375)^{12t}$$

$$t = 15.43 \text{ years}$$

$$2 = 1e^{.045t}$$

Logistic growth

(L1) Any exponentially-growing population eventually exceeds the limits (physical or biological) of its environment.

(L2) An exponentially-growing population may die out completely, or experience cycles of extremes.

(L3) Another possible population growth model (perhaps more realistic) is **logistic growth**. As a population approaches an upper limit C (**carrying capacity**), growth slows.

(L4) Here is an example of exponential versus logistic growth.

(L4) Logistic growth is characterized by: The growth rate is proportional to the population itself (amount present) and to the difference between the carrying capacity and the population.

(L5) The **logistic DE**: $P' = kP(C - P)$

$P = $ the population $P' = $ the population growth rate
$C = $ the carrying capacity $k = $ a constant of proportionality

A solution to the logistic DE

(S1) The logistic is differential equation really an IVP:

$$P'(t) = kP(t)(C - P(t)); \qquad P(0) = P_0$$

$t = $ time since the original measurement; $P(t) = $ population at time t;
$P'(t) = $ population growth rate at time t

(S2) A solution: $P(t) = \dfrac{C}{1 + de^{-kCt}}$ $P' = kP(c - P)$

d is a constant that depends upon the initial population.

Example: In some southern states the fire-ant population has grown and become a dangerous nuisance. Suppose the logistic model describes the population growth in a typical fire-ant hill, the initial population is 250, and the maximum population in a fire-ant hill is 20000. After 4 days the population of fire-ants grew to 400.

(E1) Find an expression for $P(t)$, the population of fire-ants at any time t. (Solve the DE.)

$$250 = \frac{20000}{1 + de^{-k(20000)0}}$$

$$250 = \frac{20000}{1 + d}$$

$$250(1 + d) = 20000$$

$$250 + 250d = 20000$$

$$250d = 19750$$

$$d = 79.4$$

$$400 = \frac{20000}{1 + 79.4\,e^{-k\,80000}}$$

$$400 + 31600e^{80000(-k)} = 20000$$

$$31600\,e^{80000(-k)} = 19600$$

$$e^{80000(-k)} = .6203$$

$$-k = \frac{\ln .6203}{80000}$$

$$k = .00000597$$

$$P(t) = \frac{20000}{1 + 79.4\,e^{(.000006023 \cdot 20000)t}}$$

(E2) When will the hill be half full?

$$10000 = \frac{20000}{1 + 79e^{(-.1194)t}}$$

$$10000 + 790000e^{-.1194t} = 20000$$

$$790000e^{-.1194t} = 10000$$

$$e^{-.1194t} = .0127$$

$$t = \frac{\ln .0127}{-.1194} \qquad t = 36.57 \, days$$

(E3) When does the population grow fastest? What is the population then?

$$p' = kP(c - P)$$

$$36.57 \, days \qquad 10000$$

(E4) Carefully sketch a graph of $P(t)$.

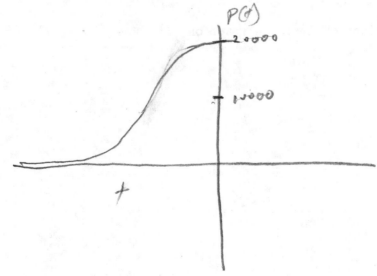

4.3   Linear and quadratic approximation; Taylor polynomials

Introduction

(I1) Use polynomial functions to approximate other (non-polynomial) functions. (Remember the polynomial approximation of a trigonometric function in Chapter 1.)

(I2) Some questions to answer:

(Q1) What do we really mean by f approximates g?

(Q2) How do we find *good* approximations?

(Q3) What role do derivatives play?

(I3) We will look at linear and quadratic approximations. Then we will work towards finding polynomial approximations of any degree: **Taylor polynomials**.

(I4) Polynomial functions are *nice*, i.e., continuous and smooth. They are ideal for approximating nasty, complicated functions.

The tangent line to the graph of f at $x = x_0$

(T1) Geometrically: the tangent line passes through the point $(x_0, f(x_0))$ and points in the direction of f; it is the straight line that best *fits* the graph of f at $x = x_0$.

(T2) Analytically: the equation of the tangent line is a linear function that best *approximates* f near $x = x_0$.

Example: Let $f(x) = \sqrt{5 - x}$.

(E1) Find an equation of the tangent line to the graph of f at $x = 4$.

Here is a graph of f and the tangent line.

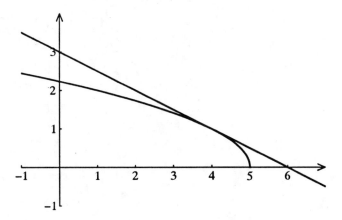

(E2) The tangent line is a good approximation to f near $x = 4$.

(E3) The graph of f is below the tangent line. (Why?)
So, the tangent line overestimates f for $x \neq 4$. ◻

Remarks:

(R1) From the previous example it appears the tangent line at $x = x_0$ is a good **linear approximation** to f near x_0.

(R2) **Constant** and **quadratic approximations** are also possible. △

Example: Which constant function best approximates $f(x) = \sqrt{5 - x}$ near $x = 4$? Which quadratic function?

Example (continued):

Here is a graph of f, and the constant, linear, and quadratic approximations to $f(x) = \sqrt{5-x}$. Which is which?

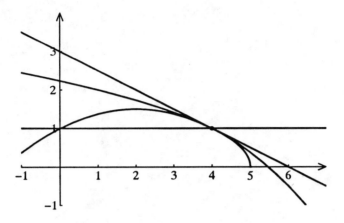

$\square$

Remarks:

(R1) All three approximating functions agree with f at $x = 4$.

(R2) The linear and the quadratic approximations are tangent to the f-graph at $x = 4$.

(R3) It appears the quadratic approximation is *better* than the linear approximation near $x = 4$.

(R4) A table of values would provide numerical evidence to support the graphs above. Of the three approximations, the quadratic does the best. $\triangle$

Here is a formal definition to summarize the technique used above to find linear and quadratic approximations.

Definition: Let f be any function for which $f'(x_0)$ and $f''(x_0)$ exist.
The **linear approximation to** f, **based at** x_0, is the linear function

$$l(x) = f(x_0) + f'(x_0)(x - x_0)$$

The **quadratic approximation to** f, **based at** x_0, is the quadratic function

$$q(x) = f(x_0) + f'(x_0)(x - x_0) + \frac{f''(x_0)}{2}(x - x_0)^2$$

Example: Let $f(x) = \log_2 x$. Find l and q, the linear and quadratic approximations to f, based at $x = 8$. Use each to estimate $f(7.9)$.

Here is a graph of f, l, and q.

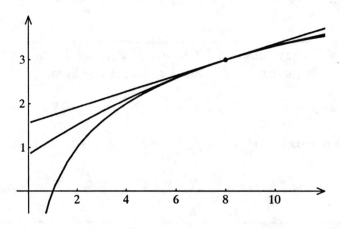

Remarks:

(R1) It seems reasonable that an even better approximation would involve a polynomial with degree 3, or maybe 4.

(R2) There is a pattern here, and it involves the derivatives of the original function. △

Definition: (Taylor polynomials.) Let f be any function whose first n derivatives exist at $x = x_0$. The **Taylor polynomial of order n, based at x_0,** is defined by

$$p_n(x) = f(x_0) + f'(x_0)(x - x_0) + \frac{f''(x_0)}{2!}(x - x_0)^2 + \frac{f^{(3)}(x_0)}{3!}(x - x_0)^3 + \cdots + \frac{f^{(n)}(x_0)}{n!}(x - x_0)^n$$

Note!

(N1) $i!$: i **factorial**

$$i! = i \cdot (i - 1) \cdot (i - 2) \cdots 2 \cdot 1 \qquad \text{and} \qquad 0! = 1$$

(N2) Constant approximation: zeroth-order approximation to f near x_0.

Linear approximation: first-order approximation to f near x_0.

Quadratic approximation: second-order approximation to f near x_0. ◇

Example: Find the nth-degree Taylor polynomial, p_n, for $f(x) = e^x$ based at $x_0 = 0$.

□

Example: Find the 5th-degree Taylor polynomial, p_5, for $f(x) = \dfrac{1}{x+1}$ based at $x_0 = 0$.

Example: Find the 4th-degree Taylor polynomial, p_4, for $f(x) = \sqrt{5-x}$ based at $x_0 = 4$. (Hint: We already found p_3 in an earlier example.)

$\square$

Remarks:

(R1) Taylor polynomials based at $x_0 = 0$ are called **Maclaurin polynomials**. So a Maclaurin polynomial for f looks like this:

$$P_n(x) = f(0) + f'(0)x + \frac{f''(0)}{2!}x^2 + \cdots + \frac{f^{(n)}(0)}{n!}x^n$$

(R2) For a Taylor polynomial p_n of order n : $p_n(x_0) = f(x_0)$ and the first n derivatives of p_n at $x = x_0$ agree with those of f.

(R3) p_n has degree at most n.

(R4) There are lots of *symbols* in the definition of the Taylor polynomial: a, x, i, x_0, f, and n. But x is the only *variable*.

(R5) From p_n to p_{n+1} : $p_{n+1}(x) = p_n(x) + \dfrac{f^{(n+1)}(x_0)}{(x+1)!}(x - x_0)^{n+1}$ $\triangle$

Example: Find P_2, P_4, and P_6, the Maclaurin polynomials, for $f(x) = \cos x$.

Here is a graph of f, P_2, P_4, and P_6.

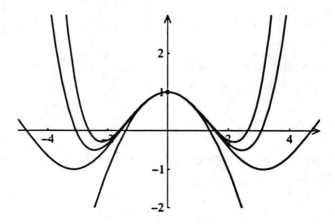

□

> **Theorem 2.** (An error bound for linear approximation.) Suppose that the inequality $|f''(x)| \leq K$ holds for all x in I. Then for all x in I,
>
> $$|f(x) - l(x)| \leq \frac{K}{2}(x - x_0)^2$$

Remarks:

(R1) $e(x) = f(x) - l(x)$: an **error function.**

   $e(x)$ is the error committed by l in approximating f. Hopefully $e(x) \approx 0$

(R2) $|f(x) - l(x)| \leq \frac{K}{2}(x - x_0)^2$: an **error bound formula.**

(R3) *Any* upper bound for $|f''(x)|$ on I will do. But finding a K as small as possible makes the error bound look better.

(R4) Suppose that the inequality $|f^{(n+1)}(x)| \leq K$ holds for all x in I.

   Then for all x in I, $\quad |f(x) - p_n(x)| \leq \frac{K}{(n+1)!}(x - x_0)^{n+1}$ $\qquad\qquad \triangle$

Example: Let $f(x) = \sin(2x)$.

(E1) Find the linear approximation, l, to $f(x)$ at $x = 0$ (the first-order Maclaurin polynomial).

(E2) What accuracy does the error bound formula guarantee if l approximates f on the interval $[-\pi/4, \pi/3]$?

(E3) Find the third-degree Maclaurin polynomial, P_3. What does the general error formula guarantee for the interval $[-\pi/2, \pi/2]$?

□

4.4 Newton's method: finding roots

Introduction

(I1) Suppose f is a differentiable function. Let r be a value in the domain of f such that $f(r) = 0$. Then r is a **root** of the equation $f(x) = 0$.

(I2) A root of f is simply an x-intercept of the f-graph. It's pretty easy to approximate a root by zooming in. Most calculators even have a root finder!

(I3) Sometimes we can find roots algebraically. If not, then Newton's method offers an alternative. And it's based on the tangent line!

Outline of Newton's method

(N1) Let x_0 be in the domain of f and an initial guess for r.

Find the tangent line (linear) approximation of f based at $x = x_0$.

$$l_1(x) = f(x_0) + f'(x_0)(x - x_0)$$

$$f(r) = 0 \approx f(x_0) + f'(x_0)(r - x_0)$$

$$r \approx x_0 - \frac{f(x_0)}{f'(x_0)} \qquad \text{provided} \qquad f'(x_0) \neq 0$$

Note! $f'(x_0) \neq 0$: the tangent line to $y = f(x)$ at $(x_0, f(x_0))$ is not horizontal, so this tangent line must intersect the x axis for a unique value x. $\diamond$

(N2) Let x_1 denote the root of the equation $l_1(x) = 0$.

Find the linear approximation of f based at $x = x_1$.

$$l_2(x) = f(x_1) + f'(x_1)(x - x_1)$$

$$f(r) = 0 \approx f(x_1) + f'(x_1)(r - x_1)$$

$$r \approx x_2 = x_1 - \frac{f(x_1)}{f'(x_1)} \qquad \text{provided} \qquad f'(x_1) \neq 0$$

(N3) Continue in this manner:

$$r \approx x_3 = x_2 - \frac{f(x_2)}{f'(x_2)} \quad \text{where } x_2 \text{ is a root of } l_2(x) = 0.$$

(N4) Each approximation is obtained from the previous approximation in the same way.

x_{n+1} is the x-intercept of the tangent line at $x = x_n$.

Here is a graph.

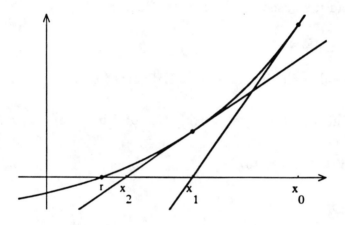

Fact: (Newton's iteration formula.) Let f be a function and x_0 an initial guess at a root. For $n \geq 0$,

$$x_{n+1} = x_n - \frac{f(x_n)}{f'(x_n)}$$

Remarks:

(R1) $x_1, x_2, x_3, \ldots$ are defined **iteratively**.

Each term is defined by its predecessor.

Each successive use of the formula is called an **iteration**.

(R2) If x_n happens to be a root of f :

$$x_{n+1} = x_n - \frac{f(x_n)}{f'(x_n)} = x_n - \frac{0}{f'(x_n)} = x_n$$

(R3) Define the **iteration function** N by $N(x) = x - f(x)/f'(x)$.

$$x_1 = N(x_0), \quad x_2 = N(x_1), \quad x_3 = N(x_2), \quad \ldots \quad x_{n+1} = N(x_n)$$

r is a root of f if and only if $N(r) = r$.

r is called a **fixed point** of the function N.   △

Example: Let $f(x) = -3x^3 - 2x^2 - 2$. Using a graph of $y = f(x)$, estimate a root of the equation $f(x) = 0$. Use Newton's method to refine your estimate.

□

Example: Use Newton's method to find an estimate of $\sqrt[3]{7}$.

□

Example: Use Newton's method to estimate solutions to the equation $4 - x^2 = \sin x$.

□

4.5   Splines: connecting the dots

Problem: How does one draw a *curve* (any unbroken line, with or without corners) through several points in a plane?

Solution: It depends on who does the drawing! Check out these examples.

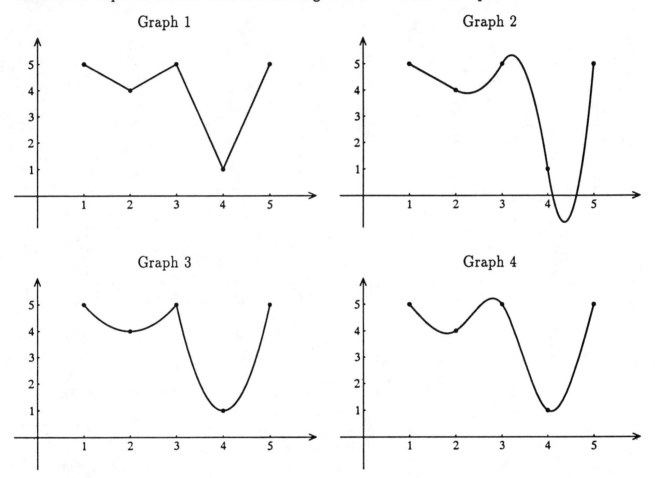

Splines

(S1) The idea is to *tie* small curve segments, called **elements**, into a single curve, called a **spline**. The *defining* points, spots where the elements are tied together, are called **knots**.

(S2) Graph 1: a **linear spline**, formed from four line segments.

Graph 2: a **quadratic spline**, formed from pieces of parabolas.

Graph 3: just another way of joining the points.

Graph 4: a **cubic spline**, formed from pieces of cubic arcs.

Example: Write a formula for the linear spline in Graph 1.

Graph 1 is given by a **piecewise-defined**, or more specifically a **piecewise-linear**, function.

Note! A linear spline is a piecewise-linear function that is also continuous.   ◇ □

Quadratic splines

(Q1) Linear splines are easy to construct, but they have those nasty corners, or sharp edges.

(Q2) Quadratic splines are smoother since they are formed by joining parabolic pieces.

(Q3) Quadratic splines are very flexible. Given two successive knots, there are lots of parabolas joining them with any desired slope at the left knot. Here is an example.

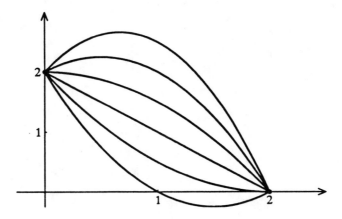

Example: Use quadratic pieces to form a smooth spline through the same knots as in the previous example:

$$(1,5), \quad (2,4), \quad (3,5), \quad (4,1), \quad (5,5)$$

Construct a **piecewise-quadratic function**.

Make the first spline element S_1 from $x = 1$ to $x = 2$ as easy as possible. That would be a straight line.

Write the second spline element S_2 in powers of $(x-2)$.

$$S_2(x) = a + b(x-2) + c(x-2)^2$$

S_2 must join the second and third points and agree in slope with S_1 at $x = 2$.

$$S_2(2) = 4, \qquad S_2'(2) = -1, \quad S_2(3) = 5$$

The remaining pieces are found in the same way.

$$S_3(x) = a + b(x-3) + c(x-3)^2; \qquad S_3(3) = 5, \quad S_3'(3) = 3, \quad S_3(4) = 1$$

$$S_4(x) = a + b(x-4) + c(x-4)^2; \qquad s_4(4) = 1, \quad S_4'(4) = -11, \quad S_5(5) = 5$$

Here is what S finally looks like.

$$S(x) = \begin{cases} S_1(x) = 5 - (x-1) & = -x + 6 & \text{if } 1 \leq x \leq 2 \\ S_2(x) = 4 - (x-2) + 2(x-2)^2 & = 2x^2 - 9x + 14 & \text{if } 2 < x \leq 3 \\ S_3(x) = 5 + 3(x-3) - 7(x-3)^2 & = -7x^2 + 45x - 67 & \text{if } 3 < x \leq 4 \\ S_4(x) = 1 - 11(x-4) + 15(x-4)^2 & = 15x^2 - 131x + 285 & \text{if } 4 < x \leq 5 \end{cases}$$

Here is a graph of S.

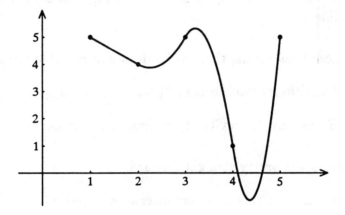

The splines fit together nicely at the knots. □

Definition: A **quadratic spline** is a continuous, **piecewise-quadratic** function S, whose first derivative S' is also continuous.

A procedure for constructing quadratic splines

Suppose $(x_0, y_0), (x_1, y_1), \ldots, (x_n, y_n)$ is a collection of knots.

(S1) Choose any quadratic curve S_1 joining (x_0, y_0) to (x_1, y_1). A straight line is fine. $m_1 = S_1'(x_1)$

(S2) Find a quadratic curve S_2 joining (x_1, y_1) to (x_2, y_2) with $S_2'(x_1) = m_1$.

$$S_2(x) = a + b(x - x_1) + c(x - x_1)^2, \qquad m_2 = S_2'(x_2)$$

(S3) Continue in this manner. For each pair of knots (x_i, y_i) and (x_{i+1}, y_{i+1}) find a quadratic curve S_{i+1} joining the knots with $S_{i+1}'(x_{i-1}) = m_i$

Remarks:

(R1) A quadratic spline has a continuous first derivative.

   But the second derivative is discontinuous at each knot.

(R2) **A cubic spline** is a piecewise-cubic function S for which the first two derivatives, S' and S'' are continuous.

(R3) At each knot, these cubic elements pieced together agree in their values and in their first two derivatives. So there are no quick changes in concavity, and this produces a smoother spline.

(R4) Even with these constraints, there are still lots of possible cubic splines.

   The **natural spline** method forces $S''(x_0) = 0 = S''(x_n)$.

   This means S has zero concavity at the first and last knots.

(R5) The calculations are pretty straight forward.

   But we'll leave most of this to a computer or calculator. $\triangle$

4.6 Optimization

Introduction

(I1) Optimization means finding maximum or minimum values of functions.

(I2) The derivative is useful for solving optimization problems: if $f'(a) = 0$ then a is a candidate for a maximum or a minimum of f.

(I3) There are some little things to worry about:

(W1) A root of the equation $f'(x) = 0$ may correspond to a local extreme value.

(W2) There may be other candidates for extreme values.

Terminology

(T1) A **stationary point** of a function f is a number x_0 such that $f'(x_0) = 0$

(T2) A **critical point** of a function f is a number x_0 such that $f'(x_0) = 0$ or $f'(x_0)$ does not exist.

Example: Find the critical points of $f(x) = 3x^4 + 2x^3 - 45x^2 + 12$.

$\square$

Example: Find the critical points of $g(x) = x^2(x-3)^{2/3}$.

$\square$

Example: Find the critical points of $h(x) = \dfrac{x^2 - 3}{x - 2}$.

$\square$

A reminder

(R1) Local maximum and minimum values of a function may or may not be global.

(R2) Here is an example of a function that has lots of local maxima and minima. Are any global?

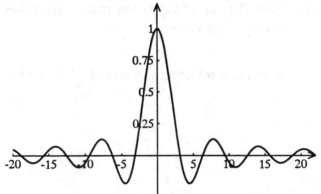

> **Fact**: A continuous function f on a closed interval $[a, b]$ can assume its maximum and minimum values *only* at critical points in (a, b) or at endpoints $[a, b]$

Remarks:

(R1) This is called the **Extreme Value Theorem**. It is an *existence* theorem. If the assumptions are met, then there must be a global minimum and a global maximum.

(R2) Where do you look?

    (1) Stationary points: roots of the equation $f'(x) = 0$

    (2) Points at which f' does not exist.

    (3) Endpoints of the interval: a and b

(R3) If the conditions of continuity and a closed interval are not met, then there is no telling what might happen. But critical points are still the key!

There may still be global extrema. Try $f(x) = |x|$, $-2 < x \leq 3$.

There may be neither a global minimum or global maximum.
Try $f(x) = 2x + 5$, $-2 < x < 3$.

There may be one without the other. Try $f(x) = 4 - x^2$, $-3 < x < 3$.   △

A procedure to find global extrema of a continuous function f on a closed interval $[a, b]$:

(S1) Find the values of f at the critical points in (a, b).

(S2) Find $f(a)$ and $f(b)$ (the values of f at the endpoints of the interval).

(S3) The largest value from (S1) and (S2) is the maximum value, the smallest value from (S1) and (S2) is the minimum value.

Example: Find the maximum and minimum values of $f(x) = x^4 - \dfrac{9}{2}x^2$ on $[-2, 3]$.

$\square$

Example: Find the maximum and minimum values of $g(x) = \cos(2x) + 2\cos(x)$ on $[0, 2\pi]$.

□

Some optimization problem jargon

(J1) The **objective function** describes the quantity to be maximized or minimized.

(J2) The **constraint equation** describes a condition that must be satisfied by the variables in an optimization problem.

(J3) A **constrained optimization problem** involves maximizing or minimizing an objective function subject to one or more constraint equations.

Example: Two nonnegative integers sum to 300. Find the two integers so that the product of one with the square of the other is a maximum.

□

Example: Find the point on the graph of $y = 4 - x^2$ that is closest to the point $(3, 4)$.

$\square$

Example: A rectangular, shoebox-type container with open top, used for shipping, must have volume 220 in^3 and the base of the container must be a square. The cardboard for the base costs 20 cents per square inch and the material for the sides costs 15 cents per square inch. Find the dimensions of the box to minimize the cost.

□

4.7 Calculus for money: derivatives in economics

Introduction

(I1) There are lots of functions from economics that we might consider.

$W(t) =$ average hourly wage at a local steel mill at any time t

$E(t) =$ number of people working for the IRS at any time t

$C(x) =$ seller's total cost of x Christmas trees

$p(x) =$ price per Christmas tree at which x trees can be sold

(I2) W and E are functions of time. Their derivatives measure rates with respect to time.

C and p are functions of a quantity. Their derivatives measure rates of change with respect to quantity.

(I3) A variable that can be only isolated values on a number line (or a finite number of values, or countably infinite number of values) is called a **discrete variable**.

A **continuous variable** can assume any real number (in an interval).

(I4) In the real world, many functions involve only discrete variables, but we can approximate them with functions of continuous variables.

Example: Don has been operating a Christmas tree farm for a number of years, and around October he must decide how many trees to tag. These trees will be cut and sold during the holiday season. Don knows that the price per tree p (in dollars) depends on the quantity x he wants to sell. His best guess for a **price function** is

$$p(x) = 9 - \frac{x}{5000}$$

If he sells x trees at $p(x)$ dollars each, he'll receive $x \cdot p(x)$ dollars. So, the **total revenue function** is

$$R(x) = x \cdot p(x) = x \left(9 - \frac{x}{5000} \right) = 9x - \frac{x^2}{5000}$$

There are fixed costs of running the business and other costs associated with growing each additional tree. The **total cost function** is

$$C(x) = 10000 + 3.5x$$

How many trees should Don tag (and cut) and sell? What should the selling price be? How much money will he make?

Don would like to maximize **profit**.

The **total profit**, P, depends on x (the number of trees sold).

$$P(x) = R(x) - C(x) = 9x - \frac{x^2}{5000} - (10000 + 3.5x) = 5.5x - \frac{x^2}{5000} - 10000$$

Here are graphs of P, R, and C. (Which is which?)

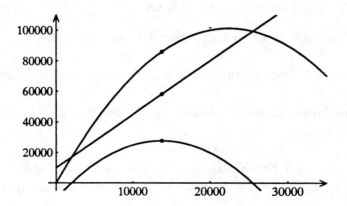

The bulleted points on the graph suggest the maximum profit occurs for $x \approx 13750$.

Find the maximum profit analytically (and how many trees should be sold, and the selling price).

Example: Take a look at R', C', and P' from the previous example. (Which is which?)

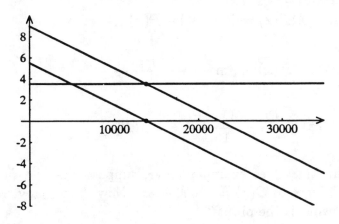

Once again something interesting occurs at $x = 13750$.

$$C'(13750) = R'(13750) = 3.5; \qquad P'(13750) = R'(13750) - C'(13750) = 0$$

At $x = 13750$, c and R increase at the same rate; P is stationary. $\qquad \square$

Terminology

(T1) $P'(7500) = 2.5$

> After selling 7500 trees, the seller is profiting at a rate of $2.50 per tree (to the mathematician).

> After selling 7500 trees, the seller makes $2.50 profit on the *next tree* (to the economist).

(T2) **Marginal profit at x :** the additional profit on the *next*, or $x + 1$st, item.

> $P'(13750) = 0$: there is 0 profit on the next item.

(T3) **Marginal cost:** the additional cost on the next item.

> **Marginal revenue:** the additional revenue on the next item.

Note! The *rate* and *marginal* interpretations of the derivative are really pretty close.

Marginal profit: $MP(x) = P(x+1) - P(x)$

Derivative $P'(x)$: $P'(x) = \lim_{h \to 0} \dfrac{P(x+h) - P(x)}{h}$

Let $h = 1$: $P'(x) \approx \dfrac{P(x+1) - P(x)}{1} = MP(x)$ ◇

Example: In the Christmas tree example above, suppose the cost function is affected by fertilizer costs and changes to $C(x) = 10000 + 4x$. Now, how many trees should Don sell? At what price? And what is the profit?

◻

Example: Try the Christmas tree problem given above with a more general cost function $C(x) = 10000 + t \cdot x$. How many trees should Don sell? At what price? And what is the profit?

□

4.8   Related rates

Introduction

(I1) Suppose $y = f(t)$ where y is a quantity that depends upon time t.

(I2) $\dfrac{dy}{dt} = f'(t)$ represents the instantaneous rate of change of y with respect to t (time).

(I3) Suppose two or more quantities that depend upon t are related by an equation. The **related rate** expression can be obtained by differentiating both sides of the equation with respect to t.

Procedure for solving related rates problems

(S1) Determine an equation relating two or more quantities that vary with time.

(S2) Differentiate both sides with respect to time, using the chain rule.

(S3) The resulting expression relates the rates at which the quantities vary.

(S4) Use the resulting expression to solve for the desired rate.

Example: Let A be the area of a triangle with base b and height h.
Find $\dfrac{dA}{dt}$ in terms of $\dfrac{db}{dt}$ and $\dfrac{dh}{dt}$.

Example: The radius of the circle in a pond created by the ripple from a stone is increasing at the rate of 2 in/sec. At what rate is the area of the circle increasing when the radius is 2 feet? 48 inches?

Example: A spherical balloon is being inflated at the rate of 4 ft^3/min. Find the rate of change of the radius of the balloon when the radius is 18 inches. How about when the radius is 2 feet?

□

Example: Two cruise ships leave a tropical island at the same time. One travels due east at 24 k/h and the other travels due north at 18 k/h. What is the rate of change of the distance between the ships one hour later? Two hours later?

□

4.9   Parametric equations, parametric curves

Curves in the plane may be specified by equations in several forms. Here are three common forms.

(C1) y as a function of x : $y = x^2 + \sin x$

(C2) x as a function of y : $x = y^2 + 3y - 1$

(C2) An expression involving x and y : $x^2 + y^2 - 4x - 9y - 4 = 0$

Suppose a point P travels around the xy-plane, tracing its path as it goes, and its coordinates are given by $x = f(t)$ and $y = g(t)$, $t \in I$.

(P1) The variable t is called a **parameter**.

(P2) f and g are called **coordinate functions**.

(P3) The figure traced out by P is called a **parametric curve**.

(P4) The two equations $x = f(t)$ and $y = g(t)$ together are called **parametric equations**.

Remarks:

(R1) The path of the point P, or the curve, is said to be defined **parametrically**.

(R2) The **direction** of the parametric curve is determined by the direction the point $(f(t), g(t))$ moves as t increases through I.   △

Example: Suppose the coordinates of a point P are given by the parametric equations

$$x = t + \sin(2t); \quad y = 3 + 3\cos t; \quad -10 \leq t \leq 10$$

What curve does P trace out? Where is P at $t = -3$? In what direction is P moving?

The easiest way to draw the curve is to plot points (x, y), and connect the dots. Here is a partial table.

t	0	1	2	3	4	5	6
x	0	1.91	1.24	2.72	4.99	4.46	5.46
y	3	2.62	.75	.03	2.04	5.85	8.88

Here is the graph of the curve C.

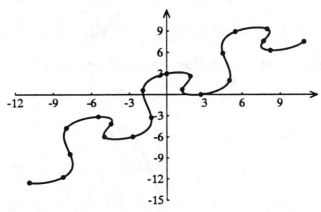

(E1) Points corresponding to integer values of t are shown with bullets.

At $t = -3$, P has coordinates $(-2.72, -5.97)$.

(E2) The curve C is *not* the graph of a function.

But certain *pieces* of C do define functions.

(E3) The graph shows the x- and y-axes, but not t-axis.

The t values usually do not appear, but some may be part of the graph.

(E4) The bullets on the graph appear at equal time intervals, but not at equal distances from each other.

P speeds up and slows down as it moves.

We'll figure out how to calculate speed of a parametric curve at a point.

(E5) If t measures time we can visualize C *dynamically*, as a curve traced by a moving point.

Curves defined by ordinary equations in x and y are *static* objects.   □

Remarks:

(R1) Parametric curves come in a wide variety, some are beautiful, some are useful, and many are interesting.

(R2) Every ordinary function graph may be written in parametric form.

Suppose $y = x^3 - 3x + 2$.

Parametrically: $x = t$; $y = t^3 - 3t + 2$ △

Example: Consider the curve

$$x = (\ln t)(\sin t); \quad y = \cos(4t); \quad 1 \le t \le 6$$

Here is the curve, with bullets indicating the initial and terminal points.

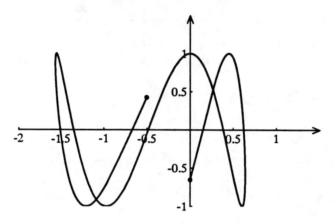

□

Example: Consider the curve

$$x = t - 2\sin t; \quad y = 1 - 3\cos t; \quad -4\pi \le t \le 4\pi$$

Here is the curve, with bullets indicating the initial and terminal points.

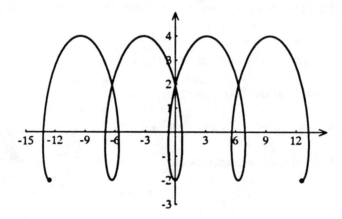

□

Remark: Parametric curves may have loops, cusps, vertical tangents, and lots of other peculiar features.   △

Example: The simplest parametric description of the unit circle is:

$$x = \cos t; \quad y = \sin t; \quad 0 \le t \le 2\pi$$

Here is the graph, with integer multiples of $\pi/2$ bulleted.

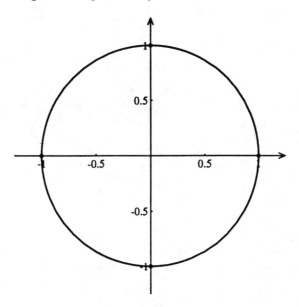

(E1) The parameter t may be thought of as time.

t may also be thought of as the radian measure of the angle determined by the x-axis and the line from the origin to P.

Then, for any angle t, $(x, y) = (\cos t, \sin t)$ is the point on the unit circle lying t radians counterclockwise from $(1, 0)$

(E2) $x^2 + y^2 = \cos^2 t + \sin^2 t = 1$

Reducing two parametric equations to xy-form is called **eliminating the parameter**.

(E3) The circle is traced once as t goes from 0 to 2π.

For a larger t-interval, the circle would be traced repeatedly. ☐

Remark: If (a, b) is any point in the plane, and $r > 0$, then the parametric equations

$$x = a + r \cos t; \quad y = b + r \sin t; \quad 0 \le t \le 2\pi$$

produce a circle with center (a, b) and radius r. △

Example: There are lots of other curves with periodic coordinate functions. Consider the curve

$$x = \sin(2t); \quad y = \cos(3t); \quad 0 \le t \le 2\pi$$

Here is a graph. Where does the curve start? End?

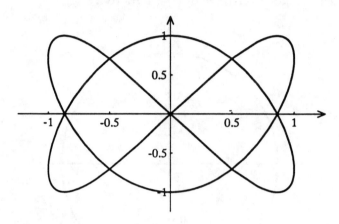

□

Note! Different parametric equations may produce the same curve in the xy-plane. Consider

(N1) $x = t + 1; \quad y = t^2; \quad -1 \le t \le 1$ and $x = \cos t + 1; \quad y = \cos^2 t; \quad 0 \le t \le \pi$

(N2) $x = t^2 + 1; \quad y = t^4; \quad -1 \le t \le 1$ and $x = \sin t + 1; \quad y = \sin^2 t; \quad 0 \le t \le \pi$

Curves may be geometrically identical and differ only in how they are traced out. ◇

Definition: Suppose that the position of a point P at time t, $a \le t \le b$, is given by differentiable coordinate functions $x = f(t)$ and $y = g(t)$. Then

$$\textbf{speed of } P \textbf{ at time } t = \sqrt{f'(t)^2 + g'(t)^2}$$

Remarks:

(R1) Showing that this definition is reasonable involves a sneaky definition of arclength.

(R2) If the position of a point P is given by *linear* coordinate functions, then we can compute the speed and see that the definition works.

     For example, let $x = at + b; \quad y = ct + d; \quad a, b, c,$ and d are constants.   △

Example: Consider the parametric curve $x = f(t) = t - \sin t$; $y = g(t) = 1 - \cos t$. Find the speed at $t = 2$. When does P move fastest?

□

Definition: The parametric curve C defined by

$$x = f(t); \quad y = g(t); \quad a \le t \le b$$

is **smooth** if f' and g' are continuous functions of t, and f' and g' are not simultaneously zero.

Theorem 3. Let the smooth parametric curve C be defined as above. If $f'(t) \ne 0$, then the slope dy/dx at the point $(x, y) = (f(t), g(t))$ is given by

$$\frac{dy}{dx} = \frac{g'(t)}{f'(t)} = \frac{dy/dt}{dx/dt}$$

Example: Consider the parametric curve given by $x = f(t) = t - \sin t$; $y = g(t) = 1 - \cos t$. Find the slope at $t = \pi/4$. Where is the curve horizontal? Where is it vertical? Carefully sketch the curve C.

□

Example: A major league pitcher throws a fastball horizontally with initial speed s_0 and initial height 7 feet.

(E1) Ignoring wind resistance, describe the ball's trajectory.

(E2) When does the ball cross home plate, 60.5 feet away?

(E3) Plot a few trajectories for various values of s_0.

(E4) Suppose wind resistance does affect the ball. How would the trajectory change?

□

4.10   Why continuity matters

Recall:

(R1) f is continuous at $x = a$ if $f(x) \to f(a)$ as $x \to a$.

(R2) The standard elementary functions, algebraic, exponential, logarithm, and trigonometric functions, are continuous wherever they are defined.

(R3) Sums, differences, quotients, and compositions of elementary functions are also continuous wherever they are defined.

(R4) Continuity is a desirable property. Two important theorems follow, each requires continuity.

Theorem 4. (The intermediate value theorem.) Let f be continuous on the closed and bounded interval $[a, b]$, and let y be any number between $f(a)$ and $f(b)$. Then for some input c between a and b, $f(c) = y$.

Remarks:

(R1) Graphically:

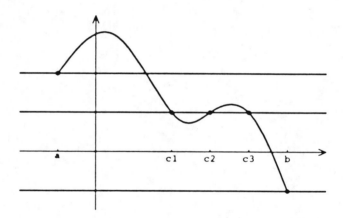

If $y = N$ is any line between $y = f(a)$ and $y = f(b)$, then $y = f(x)$ must intersect $y = N$ at least once.

Here: $f(c_1) = f(c_2) = f(c_3) = N$

(R2) If f is continuous on a closed interval, then f assumes every value between $f(a)$ and $f(b)$.

(R3) If f is not continuous, then the conclusion of the Intermediate Value Theorem may or may not hold.

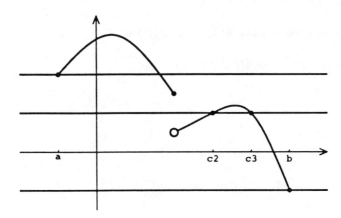

Here, $f(c_2) = f(c_3) = N$

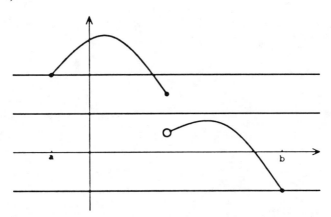

Here, there is no c such that $f(c) = N$.

(R4) The IVT is an *existence* theorem. It guarantees there is some point in the domain such that $f(c) = N$, but it does not **say where** or how many there may be. $\triangle$

The IVT is used to derive the **bisection method**, a technique for solving equations. Here is a little background.

(B1) A number r is a **root** (or **zero**) of a function f if $f(r) = 0$.

(B2) If $f(a)$ and $f(b)$ have opposite signs, then $N = 0$ is an intermediate value. So there must be at least 1 $c \in (a, b)$ such that $f(c) = 0$

Fact: If f is continuous on $[a, b]$, and $f(a)$ and $f(b)$ have opposite signs, then $[a, b]$ contains a root of f. Hence the *midpoint* $(a + b)/2$ of $[a, b]$ differs from a root of f by no more than $(b - a)/2$ - the *radius* of $[a, b]$.

Example: Consider $f(x) = x^2 - 2x - 1$

(E1) Show there is a c such that $f(c) = 2$. Find c.

(E2) Does f have a zero in $[0, 4]$. If so, find one.

□

Example: Use the bisection algorithm to estimate $\sqrt{7}$.

$\square$

Note! Each step in the bisection algorithm doubles the precision of the estimate. Suppose m_n is the n-th midpoint and c is the root. Then

$$|m_n - c| < \frac{b - a}{2^n}$$

There is no guarantee that we will find the root c exactly. But at least we can get good approximations. $\diamond$

Theorem 5. (The extreme value theorem.) Let f be continuous on the closed, bounded interval $[a, b]$. Then f assumes both a maximum value and a minimum value somewhere on $[a, b]$

Remarks:

(R1) Like the IVT, the EVT is an existence theorem.

(R2) If the function is not continuous and/or the interval is not closed, then absolute extrema are not guaranteed.

(R3) Graphically:

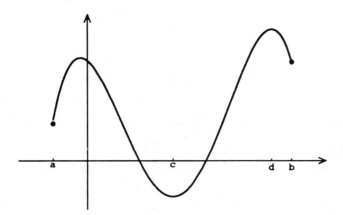

$f(c)$:   absolute minimum value

$f(d)$:   absolute maximum value

(R4) If f is continuous on $[a, b]$ then search for absolute extrema at the local extrema and the endpoints a and b.

(R5) If f is not continuous and/or the interval is not closed, there may or may not be absolute extrema in the interval. A sketch may help.   △

Example: Let $f(x) = 1/x$.

(E1) What does the EVT say about f on the interval $[.1, 1]$?

(E2) f is continuous on $[1, \infty)$, but it has no minimum value on this interval. Does this contradict the EVT?

(E3) f is continuous on $(0, 1]$, but it has no maximum value on this interval. Does this contradict the EVT?

(E4) What does the EVT say about f and the interval $[0, 1]$?

(R5) $f(-1)$ and $f(1)$ have opposite signs. Why isn't there a root in $[-1, 1]$?

Example (continued):

4.11   Why differentiability matters; the mean value theorem

Recall:

> **Definition**: Let f be a function defined near and at $x = a$. The derivative of f at $x = a$, denoted $f'(a)$ is defined by the limit
>
> $$f'(a) = \lim_{h \to 0} \frac{f(a + h) - f(a)}{h}$$

Remarks:

(R1) $f'(a)$ may *not* exist.
      If $f'(a)$ does exist, f is **differentiable at** $x = a$, or f **has a derivative at** $x = a$.

(R2) The standard elementary functions - algebraic, exponential, logarithm, trigonometric functions - are differentiable wherever they are defined.

(R3) Combinations - sums, products, quotients, compositions - of elementary functions are also differentiable where they are defined.

(R4) Differentiability and continuity are related.

> **Theorem 6.** If f is differentiable at $x = a$, then f is also continuous at $x = a$.

Proof:

□

Theorem 7. (The mean value theorem.) Suppose that f is continuous on the closed interval $[a, b]$ and differentiable on the open interval (a, b). Then for some c between a and b,

$$f'(c) = \frac{f(b) - f(a)}{b - a}.$$

Remarks:

(R1) A graphical view:

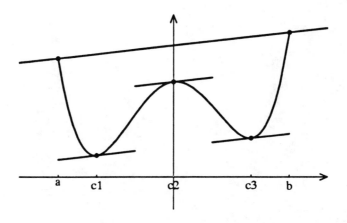

(R2) In the hypothesis: if f is differentiable on (a, b), then there are no sharp corners for $x \in (a, b)$.

If a sharp corner occurs, the conclusion of the MVT is not guaranteed.

Just take a look at $f(x) = |x|$ on $[-1, 1]$.

(R3) An interpretation:

$f'(c)$ represents the *instantaneous* rate of change at c.

$\dfrac{f(b) - f(a)}{b - a}$ represents the **average** rate of change over the interval $a \le x \le b$.

At some point c between a and b, the instantaneous rate of change is equal to the average (mean) rate of change over the entire interval. $\triangle$

Example: Consider $f(x) = 4 - 4x - x^2$, $-4 \leq x \leq 1$. Find the numbers c that satisfy the conclusion of the MVT. Sketch a graph to illustrate the conclusion of the MVT.

□

Recall: If $f(x) = k$, then $f'(x) = 0$.

The following theorem is the converse, and is a consequence of the MVT.

Theorem 8. If $f'(x) = 0$ for all x in an interval I, then f is *constant* on I.

Proof:

$\square$

Theorem 9. If $F'(x) = G'(x)$ for all x in an interval I, then $F(x) = G(x) + C$ for some constant C.

Remarks:

(R1) In words: Two functions with the same derivatives on an interval I must differ by a constant.

(R2) This theorem is proved by applying the previous theorem to the difference function $F(x) - G(x)$. $\triangle$

The MVT may be used to prove the general principle that the sign of f' determines whether f increases or decreases over an interval.

Theorem 10. Suppose that $f'(x) > 0$ for all x in an interval I. Then f is increasing on I.

Proof:

$\square$

Remark: The MVT may also be used to prove the racetrack principle. In this case, apply the MVT to the *right* function. $\triangle$

The proof of the MVT appeals to several other results.

Lemma: Suppose that f assumes either its maximum or its minimum on $[a, b]$ at a point c between a and b. If $f'(c)$ exists, then $f'(c) = 0$.

Lemma. (Rolle's theorem.) Suppose that f is continuous on $[a, b]$, differentiable on (a, b), and $f(a) = f(b)$. Then for *some* c between a and b, $f'(c) = 0$.

Remark: This is really just a special case of the MVT, $f(a) = f(b)$. $\triangle$

Example: Consider $f(x) = x^3 - 9x^2 + 23x - 13$.

(E1) Show that f satisfies the assumptions of Rolle's theorem on the interval $[1, 5]$. Find a value of c guaranteed to exist by Rolle's theorem.

(E2) Show that f satisfies the assumptions of the MVT on the interval $[1, 6]$. Find a value of c guaranteed to exist by the MVT.

(E3) Sketch a graph illustrating (E1) and (E2).

Example: There exists a classic film strip about an Alabama State Trooper, Rolle's theorem, and the MVT. The State Trooper apparently had a good calculus course while in college and used the MVT to write a speeding ticket. Here's the idea: At 6:00 a.m. a delivery truck driver took a fare card and entered the turnpike. At 7:30 a.m. he exited the turnpike at a toll booth 108 miles down the road. The toll booth operator immediately summoned the State Trooper who issued a speeding ticket. The speed limit is 65 mph.

(E1) How does the MVT prove conclusively that the truck driver was speeding?

(E2) The fine for speeding is $75 plus $10 for each mph over the speed limit. What is the truck driver's minimum fine?

□

4.12 Chapter summary

This chapter contains a few applications of the derivative and antiderivative.

(S1) A differential equation is any equation that involves a function and its derivatives.

  (D1) To solve a DE means to find a function that satisfies the equation.

  (D2) Most differential equations have many solutions.

  (D3) An initial value problem involves a DE and an initial condition. IVP's usually have only one solution.

(S2) Approximations.

  (A1) Any well-behaved function f may be approximated near x_0 by the tangent line, the linear approximation to f at x_0.

  (A2) The quadratic approximation to f at x_0 is the function whose graph is the parabola that best fits f at x_0.

  (A3) Taylor polynomials offer approximations of any degree.

(S3) Newton's method is a numerical technique for finding roots. It relies upon the tangent line to f at x_0 and is usually very accurate.

(S4) Splines are constructed from pieces of simple curves to form larger curves of desired shapes. Once again, derivatives are the basic tool.

(S5) The derivative is very useful for finding extreme values, maximum and minimum values, of functions. This involves finding and classifying critical points.

(S6) The derivative may also be used in economic applications. Marginal cost, marginal revenue, and marginal profit all have interpretations in terms of the derivative.

(S7) Suppose two or more quantities that depend upon t are related by an equation. The rates at which these quantities vary are also related. The related rate expression can be obtained by differentiating both sides of the equation with respect to t.

(S8) Parametric equations lead to some wonderful plane curves, many are not graphs of functions. If the coordinate functions are differentiable in t, then calculus ideas apply; we can find slopes and speed.

(S9) The intermediate value theorem, the extreme value theorem, and the mean value theorem are all very important calculus results. They are used to prove other theorems in calculus and also lead to some nice applications.

Chapter 5

The Integral

5.1   Areas and integrals

There are two major problems in calculus:

1. The tangent line problem: solved by using derivatives.

2. The area problem.

The area problem and the integral

(A1) Find the area of the region bounded above by the graph of f, below by the x axis, on the left by the vertical line $x = a$, and on the right by the vertical line $x = b$.

Illustration:

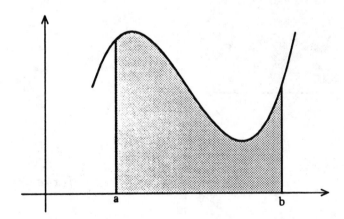

(A2) In this chapter it is convenient to talk about **signed area**.

  (S1) The area above the x-axis counts as positive.

  (S2) The area below the x-axis counts as negative.

Definition: (The integral as signed area.) Let f be a function defined for $a \leq x \leq b$. Either of the equivalent expressions

$$\int_a^b f \qquad \text{or} \qquad \int_a^b f(x)\,dx$$

denotes the signed area bounded by $x = a$, $x = b$, $y = f(x)$, and the x-axis.

Remarks:

(R1) In words: the **integral** of f from a to b. The function f is the **integrand**.

(R2) While both expressions denote the same area, the second form has some advantages.

(R3) These expressions are symbols: they stand for area. How do we calculate this area?

(R4) Sometimes the term *net* area is used to mean signed area. $\triangle$

Example: Several areas are shown below, labeled as integrals. Use familiar area formulas to evaluate each integral.

(E1)

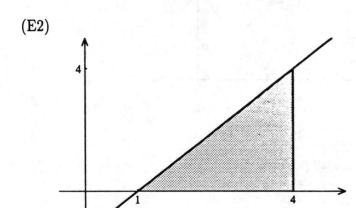

$$\int_1^5 f(x)\,dx = \quad 4 \cdot 2 = 8$$

(E2)

$$\int_1^4 f(x)\,dx = \quad \frac{1}{2} \cdot 3 \cdot 4 = 6$$

(E3)

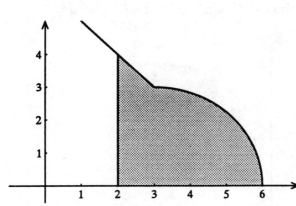

$$\int_2^6 f(x)\,dx = \quad \frac{1}{4}\tilde{\pi}(3)^2 + 3 + \frac{1}{2} = 5.72$$

$$5.22 + \# \frac{1}{2}$$

□

Example: The graph of $y = f(x)$ is shown below. Find (or estimate) values for the integrals $I_1 = \int_0^3 f(x)\,dx$ and $I_2 = \int_{-3}^3 f(x)\,dx$.

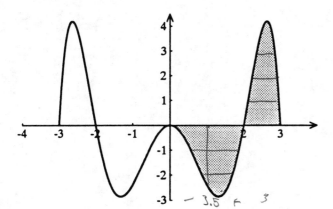

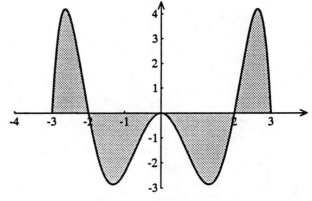

Signed area 1: $\int_0^3 f(x)\,dx$ $-3.5 + 3$

$-.5$

Signed area 2: $\int_{-3}^3 f(x)\,dx$

□

Example: Let $g(x) = x + \sin x$. Find or estimate $\int_0^\pi g(x)\,dx$ and $\int_{-\pi}^\pi g(x)\,dx$.

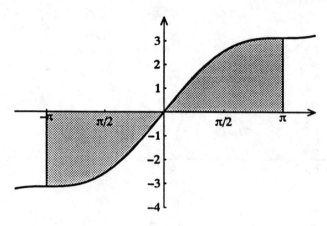

$\square$

Theorem 1. (New integrals from old.) Let f and g be continuous functions on $[a, b]$; let k denote a real constant. Then

1. $\displaystyle\int_a^b (f(x) \pm g(x))\,dx = \int_a^b f(x)\,dx \pm \int_a^b g(x)\,dx$.

2. $\displaystyle\int_a^b k f(x)\,dx = k \int_a^b f(x)\,dx$.

3. If $f(x) \le g(x)$ for all x in $[a, b]$, then $\displaystyle\int_a^b f(x)\,dx \le \int_a^b g(x)\,dx$.

4. If $a < c < b$, then $\displaystyle\int_a^b f(x)\,dx = \int_a^c f(x)\,dx + \int_c^b f(x)\,dx$.

Remarks:

(R1) The integral of a sum (difference) is the sum (difference) of the integrals.

(R2) Constants pass freely through integral signs.

(R3) Part 3 geometrically:

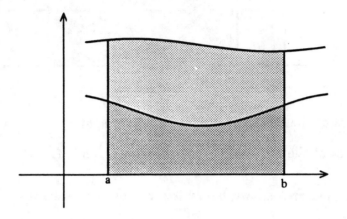

(R4) Part 4 geometrically:

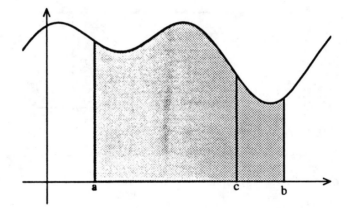

△

Property 3 is often used in the case where either f or g is a constant.

Fact: (Bounding an integral.) Suppose that for some numbers m and M, the inequality $m \le f(x) \le M$ holds for all x in $[a, b]$. Then

$$m \cdot (b - a) = \int_a^b m \, dx \le \int_a^b f(x) \, dx \le \int_a^b M \, dx = M \cdot (b - a).$$

Here is an illustration of that fact:

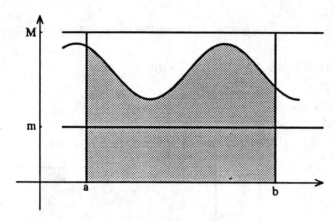

$m \cdot (b - a) =$ area of the rectangle over $[a, b]$ with height m.

$M \cdot (b - a) =$ area of the rectangle over $[a, b]$ with height M.

Example: The shaded area shown below is given by the integral $I = \int_1^3 f(x)\,dx$. Use g, h, and j to estimate I.

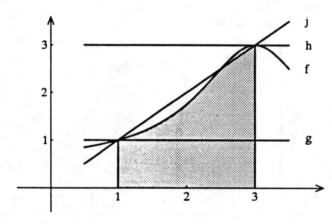

A few integral facts:

$$\int_a^b k\, dx = k(b-a) \qquad \int_a^b x\, dx = \frac{1}{2}(b^2 - a^2) \qquad \int_a^b x^2\, dx = \frac{1}{3}(b^3 - a^3)$$

$$\int_a^b \cos x\, dx = \sin b - \sin a \qquad \int_a^b \sin x\, dx = \cos a - \cos b$$

Example: Using the facts above, find the following integrals.

(E1) $\displaystyle\int_{-2}^3 (2x^2 + x)\, dx =$

(E2) $\displaystyle\int_0^\pi (3\cos x - \sin x)\, dx =$

$\square$

Average value and the integral: Consider the following graph.

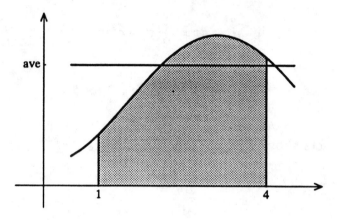

(A1) The rectangle is selected so that it has the same area as the shaded region.

(A2) The height of the rectangle is (naturally) the average value of the function f over the interval $[a, b]$.

Definition: Let f be defined on an interval $[a, b]$. The quantity

$$\frac{\int_{a}^{b} f(x)\, dx}{b - a}$$

is the **average value of f over $[a, b]$**.

Remarks:

(R1) Interpreting the integral:

    (I1) $\int_{a}^{b} f(x)\, dx$: the signed area bounded by $x = a$, $x = b$, $y = f(x)$, and the x-axis.

    (I2) Suppose $f(t)$ represents the speed of a moving object at time t.

      $\int_{a}^{b} f(t)\, dt$: the distance traveled by the object over the interval $a \leq t \leq b$.

    (I3) Suppose $v(t)$ is the velocity of an object moving along a line at time t.

      $\int_{a}^{b} v(t)\, dt$ represents the net distance traveled over the interval $a \leq t \leq b$.

(R2) It is natural to think of x as moving left to right.

In $\int_a^b f(x)\,dx$ it is reasonable to think $a \le b$.

But, *right-to-left* integrals do arise.

In general: $\int_a^b f(x)\,dx = -\int_b^a f(x)\,dx$

For example: $\int_{-1}^2 (x^3 + 2x)\,dx = -\int_2^{-1} (x^3 + 2x)\,dx$

(R3) $\int_a^b f(x)\,dx$ represents signed area.

We need a practical and efficient method for calculating signed area.

The key result for calculating signed area is the connection between the derivative and the integral: the **fundamental theorem of calculus**.

(R4) Given a function $f(x)$. Does $\int_a^b f(x)\,dx$ always exist?

In words: Does signed area always exist? $\triangle$

5.2   The area function

Introduction

(I1) Recall: $\displaystyle\int_a^b f(x)\,dx = $ signed area, a fixed number.

(I2) Now: define $A_f = $ the **area function** (or area-so-far function).

As x varies, the endpoint of the region varies.

Definition: Let f be a function and a any point of its domain. For any input x, the **area function** A_f is defined by the rule

$$A_f(x) = \int_a^x f(t)\,dt = \text{the signed area defined by } f, \text{ from } a \text{ to } x.$$

Remarks:

(R1) Here is a picture to illustrate the definition.

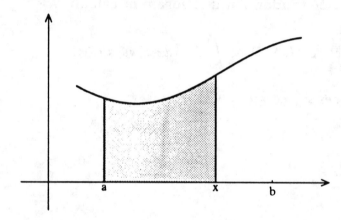

(R2) Use of t :   a *dummy* variable, don't use x in two different ways.

(R3) Domain of A_f :   same domain as f (as long as there are no discontinuities).

Even if x is to the left of a : $\displaystyle\int_a^x f(t)\,dt = -\int_x^a f(t)\,dt$

(R4) The role of a is to fix the left edge of the region. The other edge varies freely.

The choice of a only slightly affects A_f.   △

Example: Let $f(x) = 7$ and $a = 0$.

(E1) Describe the area function $A_f = \int_0^x f(t)\, dt$ for positive inputs x.

(E2) Find a formula for A_f.

(E3) Sketch f and A_f on the same coordinate axes.

(E4) Describe the area function $A_f = \int_0^x f(t)\, dt$ for negative inputs x.

Does the formula in (E2) still work?

$\square$

Example: Let $f(x) = 2x + 1$. Find a formula for $A_f(x) = \int_0^x f(t)\, dt$.

□

Example: Let $f(x) = 2x + 1$. Find a formula for $A_f(x) = \int_1^x f(t)\, dt$.

□

Note! In every example above, direct calculation showed the area function A_f to be an antiderivative of f.

Theorem 2. (The fundamental theorem of calculus, informal version.) For *any* well-behaved function f and any "base-point" a, A_f is an antiderivative of f.

Properties of A_f.

Let f be a continuous function, a a point of its domain, and $A_f(x) = \int_a^x f(t)\, dt$. Then:

(P1) $A_f(a) = 0$

(P2) Where f is positive, A_f is increasing.

(P3) Where f is negative, A_f is decreasing.

(P4) Where f is zero, A_f has a stationary point.

(P5) Where f is increasing, A_f is concave up.

(P6) Where f is decreasing, A_f is concave down.

Remarks:

(R1) Examples in graphical form help to justify each claim.

(R2) Think about the relationship between a function and its derivative in order to help justify each claim.   △

5.3   The fundamental theorem of calculus

In the last section we defined and studied the area function A_f. All the evidence seems to suggest that A_f is an antiderivative of f. The formal statement of this fact is *the most important idea in all of calculus.*

Theorem 3. (The fundamental theorem of calculus, formal version.) Let f be a continuous function, defined on an open interval I containing a. The function A_f with rule

$$A_f(x) = \int_a^x f(t)\, dt$$

is defined for every x in I, and $\dfrac{d}{dx}(A_f(x)) = f(x)$.

Remarks:

(R1) Every continuous function has an antiderivative. If f is continuous, then A_f exists.

(R2) Working with an open interval is a technical convenience. This way we avoid troubles at endpoints.

(R3) Other notation: $\dfrac{d}{dx}\left[\int_a^x f(t)\, dt\right] = f(x)$ △

Example: Let $f(x) = xe^{-x^2}$ and let $A_f(x) = \int_0^x f(t)\, dt$. Find a symbolic formula for A_f; interpret results graphically.

Here is a graph of $f(x) = xe^{-x^2}$. $A_f(1)$ is the net shaded area.

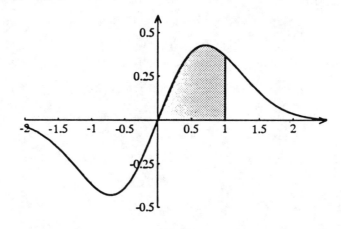

Example (continued):

Here is a graph of f and A_f.

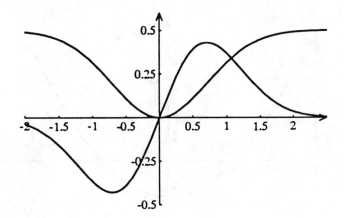

Example: Let $f(x) = 3x^3 - 3$ and let $A_f(x) = \int_1^x f(t)\, dt$ (watch that lower bound). Find a symbolic formula for $A_f(x)$. Add a sketch of A_f to the graph of f below.

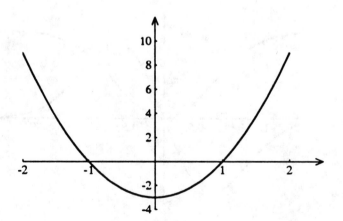

Recall: If two functions have the same derivative, then they differ by a constant. If two curves have the same *slope* then they must be *parallel.* Here is a more formal statement of this result.

Fact: Suppose that $F'(x) = G'(x)$ for all x in an interval I. Then for some constant C, $F(x) = G(x) + C$ for all x in I.

Remarks:

(R1) This fact means, if you find *one* antiderivative, you have found them all!

(R2) The FTC, part 2, uses this result to tie derivatives and integrals together, and gives us a practical method to compute certain integrals (to find area *exactly*). $\triangle$

Theorem 4. (Fundamental theorem, second version.) Let f be continuous on $[a, b]$, and let F be *any* antiderivative of f. Then

$$\int_a^b f(x)\, dx = F(b) - F(a).$$

Proof:

$\square$

Notation: $\displaystyle\int_a^b f(x)\, dx = \left[F(x)\right]_a^b = F(x)\Big]_a^b = F(b) - F(a)$

Example: Evaluate each integral and interpret as signed area.

(E1) $\int_{-2}^{5} \dfrac{x}{3}\, dx$

(E2) $\int_{0}^{3} \left(x^2 - x\right) dx$

(E3) $\displaystyle\int_0^\pi \sin x \, dx$

(E4) $\displaystyle\int_0^5 e^{-x} \, dx$

(E5) $\displaystyle\int_{-1}^{1} \frac{1}{x^2}\, dx$

(E6) $\displaystyle\int_{1}^{b} \frac{1}{x}\, dx$

□

Note!

(N1) The second version of the fundamental theorem may be restated:

Fact: Let f be a well-behaved function on $[a, b]$, with derivative f'. Then

$$\int_a^b f'(t)\,dt = f(b) - f(a)$$

In words: Integrating f' (the rate function) over $[a, b]$ gives the change in f (the amount function) over the same interval.

(N2) The FTC is very useful, but it won't work on many (simple) problems.

$$\int_0^{\pi/4} \tan(x^2)\,dx\,; \qquad \int_{-2}^2 e^{x^2}\,dx\,; \qquad \int_0^1 \sqrt{1+x^3}\,dx$$

These integrals still make good geometric sense.

(N3) Notation:

$$\int f(x)\,dx \quad (= F(x) + C) \quad \textbf{indefinite integral notation}$$

Denotes an entire family of functions (antiderivatives), one for each value of C.

$$\int_a^b f(x)\,dx \quad \textbf{definite integral}$$

This represents a *number*, the signed area.

(N4) The idea of the proof of the FTC is given in the text. ◇

5.4   Approximating sums: the integral as a limit

Introduction

(I1) The FTC is a wonderful result, but many functions don't have an easy antiderivative.

(I2) In this section we will consider another approach to finding a definite integral: a limit of approximating sums.

(I3) The formal definition has all kinds of notation, terminology, and symbols, but the basic idea is reasonable and straightforward.

Example: Part of the graph of $f(x) = -x^2 + 5$ appears below. Use the FTC to find the shaded area exactly. Then use approximating sums to estimate the same area.

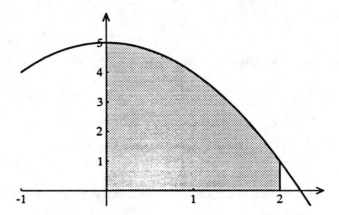

With the FTC, the exact answer is easy to find by antidifferentiation.

How could we estimate this area? The following figures suggest four possible strategies. In each case we will approximate the desired area by a sum of simpler areas, rectangles or trapezoids. Adding up the area of these simpler figures gives, in each case, a natural estimate to our desired area.

(E1) Right approximating sum with 5 subdivisions.
The height of each rectangle is the value of f at the right endpoint of each subinterval.

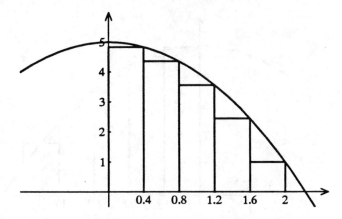

$$R_5 = f(.4)(.4) + f(.8)(.4) + f(1.2)(.4) + f(1.6)(.4) + f(2)(.4)$$

$$= (4.84)(.4) + (4.36)(.4) + (3.56)(.4) + (2.44)(.4) + (1)(.4) = 6.48$$

Because each of the five rectangles lies inside the parabolic region, one can conclude that the area of the region is greater than 6.48.

(E2) Left approximating sum with 5 subdivisions.
The height of each rectangle is the value of f at the left endpoint of each subinterval.

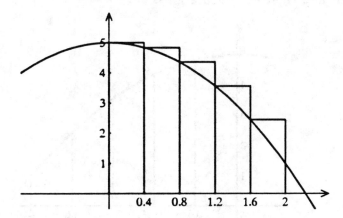

$$L_5 = f(0)(.4) + f(.4)(.4) + f(.8)(.4) + f(1.2)(.4) + f(1.6)(.4)$$

$$= (5)(.4) + (4.84)(.4) + (4.36)(.4) + (3.56)(.4) + (2.44)(.4) = 8.08$$

Because the parabolic region lies within the union of the five rectangular regions, one can conclude that the area of the region is less than 8.08.

(E3) Midpoint approximating sum with 5 subdivisions.
The height of each rectangle is the value of f at the midpoint of each subinterval.

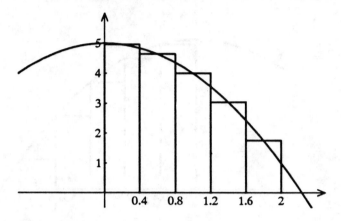

$$M_5 = f(.2)(.4) + f(.6)(.4) + f(1)(.4) + f(1.4)(.4) + f(1.8)(.4)$$

$$= (4.96)(.4) + (4.64)(.4) + (4)(.4) + (3.04)(.4) + (1.76)(.4) = 7.36$$

(E4) Trapezoid approximating sum.
Using this method looks, geometrically, like a good idea.

Fact: The trapezoid approximation with n subdivisions is the average of the corresponding left and right approximations.

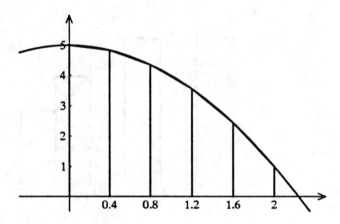

$$T_5 = (1/2)(R_5 + L_5) = (1/2)(6.48 + 8.08) = 7.28 \qquad \square$$

Sigma notation

Let m and n be positive integers, $m \le n$, and $a_m, a_{m+1}, \ldots, a_n$ a set of real numbers.

$$\sum_{i=m}^{n} a_i = a_m + a_{m+1} + a_{m+2} + \cdots + a_n$$

(S1) i : index of summation ; m : initial value ; n : final value

(S2) $\sum$: carry out the addition.

(S3) a_i : the term to be added.

(S4) $\displaystyle\sum_{i=1}^{n} a_i = \sum_{j=1}^{n} a_j = \sum_{k=1}^{n} a_k$ the index is a dummy variable.

Example: Evaluate the following summations.

(E1) $\displaystyle\sum_{i=1}^{6} \frac{i^2 - 1}{i^2 + 1} =$

(E2) $\displaystyle\sum_{i=1}^{n} \frac{1}{i} =$

(E3) $\displaystyle\sum_{j=1}^{7} \pi =$

(E4) $\displaystyle\sum_{k=1}^{n} k =$

□

Example: Use sigma notation to rewrite the left, right, and midpoint approximating sums (L_5, R_5, and M_5) from above.

The ordered set of endpoints is called a **partition** of the x-interval $[0, 2]$.

$0 < .4 < .8 < 1.2 < 1.6 < 2$ partition

$x_0 < x_1 < x_2 < x_3 < x_4 < x_5$ name the endpoints

□

Remarks:

(R1) L_5, R_5, and M_5 are approximations to the exact area under the curve.

(R2) All of the approximations get better as the number of subintervals increases.

(R3) The left, right, and midpoint approximating sums are all special types of **Riemann sums**.

(R4) A general Riemann sum:

(S1) Allows for rectangles with unequal bases. Each subinterval may have any length.

(S2) The height of each rectangle is determined by *any* number in the subinterval.

Here is a graphical version of a Riemann sum.

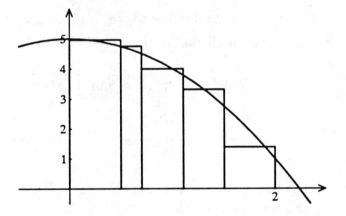

$\triangle$

Definition: Let I be partitioned into n subintervals by any $n + 1$ points

$$a = x_0 < x_1 < x_2 < \cdots < x_{n-1} < x_n = b;$$

let $\Delta x_i = x_i - x_{i-1}$ denote the width of the i-th subinterval. Within each subinterval $[x_{i-1}, x_i]$, choose any point c_i. The sum

$$S_n = \sum_{i=1}^{n} f(c_i)\, \Delta x_i = f(c_1)\, \Delta x_1 + f(c_2)\, \Delta x_2 + \cdots + f(c_n)\, \Delta x_n$$

is a **Riemann sum with n subdivisions** for f on $[a, b]$.

Remarks:

(R1) L_n, R_n, and M_n are Riemann sums.

Each is constructed using a **regular partition** of $[a, b]$;
The subintervals are of equal length.

(R2) c_i : called **evaluation points**.

(R3) A trapezoid approximating sum, T_n, is *not* a Riemann sum. The approximating figures
are not rectangles. But, a trapezoidal sum is still a good approximation to the signed
area.   △

Note! Graphical and numerical evidence suggests that the approximating sums (L_n, R_n,
M_n, T_n) should *converge* toward some number. That number should be the true signed area
or the definite integral.   ◇

Definition: Let the function f be defined on the interval $I = [a, b]$. The **integral of f**
over I, denoted $\int_a^b f(x)\, dx$, is the number to which all Riemann sums S_n tend as n tends
to infinity and as the widths of all subdivisions tend to zero. In symbols:

$$\int_a^b f(x)\, dx = \lim_{n \to \infty} S_n = \lim_{n \to \infty} \sum_{i=1}^n f(c_i)\, \Delta x_i \,,$$

if the limit exists.

Remarks:

(R1) If the limit exists then f is called **integrable** on $[a, b]$.
Almost every calculus-style function that is defined and bounded on $[a, b]$ is integrable.

(R2) There is a lot going on with this limit: x_is, c_is, ...

Fortunately, almost any reasonable approximating sum approaches the true value of
the integral as $n \to \infty$.

(R3) Consider a right approximating sum with n equal subdivisions, $(b - a)/n = \Delta x$

$$\int_a^b f(x)\, dx = \lim_{n \to \infty} \sum_{i=1}^n f(x_i)\, \Delta x$$

The dx on the left corresponds in a natural way to Δx on the right.

(R4) Calculating these sums by hands is tedious. This is a task computers and calculators
do well.   △

5.5 Approximating sums: interpretations and applications

Introduction

(I1) We have defined the integral as a limit of approximating sums.

$$\int_a^b f(x)\,dx = \lim_{n\to\infty} \sum_{i=1}^n f(c_i)\,\Delta x_i$$

(I2) In this section we will look at approximating sums more closely and apply this idea to several problems of measurement.

Approximating sums are **discrete** approximations to integrals of **continuously** varying functions. Here are three interpretations.

(I1) Geometric: The area of simpler figures is used to approximate the area of a more complex *curvy* region.

The shaded region below is a **polygonal approximation** to the area from a to b under the graph of f.

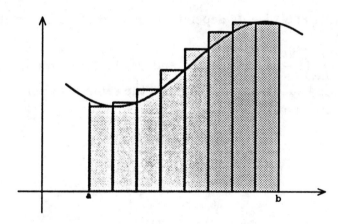

(I2) Simpler functions: In order to find an approximating sum we replace the original function f with a new, simpler function, one that is linear on each subinterval.

In the graph below, the original function f (from (I1)) is replaced by an 8-step linear approximation.

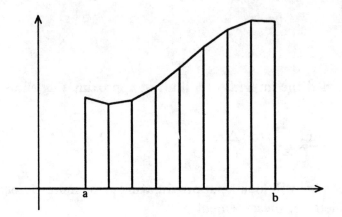

(I3) Weighted sums: A Riemann sum for $\int_a^b f(x)\,dx$ has the form

$$\sum_{i=1}^{n} f(c_i)\,\Delta x_i = f(c_1)\,\Delta x_1 + f(c_2)\,\Delta x_2 + \cdots + f(c_n)\,\Delta x_n.$$

An approximating sum is a **weighted sum** of values of f at various points c_i (the **sampling points**) in the interval $[a, b]$.

Sampling points and weights depend on the type of approximating sum.

Example: Let $I = \int_{-2}^{2}(1 - x^2)\,dx$. Let L_4 and M_4 be the left and midpoint approximating sums for I, each with 4 subdivisions. What are the sampling points? What are the weights? what are the weighted sums?

Problem:

Let f and g be any continuous functions. Let R be the region bounded above by g, below by f, on the left by $x = a$, and on the right by $x = b$. Find the area of the region R.

Here is a graph to illustrate the problem.

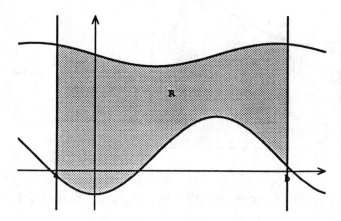

A solution:

(S1) Slice the area into vertical strips, approximate each strip by a rectangle, and add up the areas. (Find an approximating sum.)

(S2) Here is a graph to illustrate a solution.

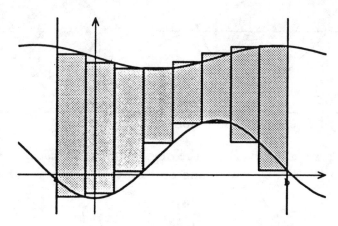

(S3) As $n \to \infty$, the approximating sum tends to the exact area between the two curves.

(S4) By the limit definition of the integral, the approximating sum also tends to an integral.

(S5) Area of $R = \displaystyle\int_a^b (g(x) - f(x))\, dx$.

Plane regions

(P1) Perhaps the simplest plane region is one bounded above by the graph of a function, below by the x-axis, on the left by $x = a$ and on the right by $x = b$.

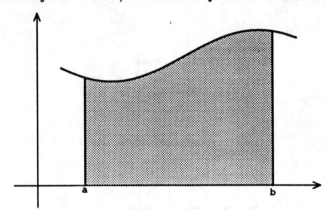

$$\text{Area} = \int_a^b f(x)\, dx$$

(P2) There are other slightly more complicated regions. Some of the curves may not be graphs of functions.

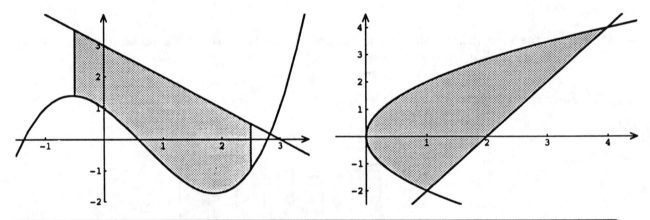

Fact: Let f and g be continuous functions.

(F1) **Integrating in x.** Let R be the region bounded *above* by $y = g(x)$, *below* by $y = f(x)$, on the *left* by $x = a$, and on the *right* by $x = b$. Then R has area

$$\int_a^b (g(x) - f(x))\, dx.$$

(F2) **Integrating in y.** Let R be the region bounded on the *right* by $x = g(y)$, on the *left* by $x = f(y)$, *below* by $y = c$, and *above* by $y = d$. Then R has area

$$\int_c^d (g(y) - f(y))\, dy.$$

Remarks:

(R1) If f and g are continuous on $[a, b]$, then $g - f$ is continuous on $[a, b]$,

and $\int_a^b (g(x) - f(x))\, dx$ exists.

(R2) $\int_a^b (g(x) - f(x))\, dx = \int_a^b (y_{\text{upper}} - y_{\text{lower}})\, dx$

$y_{\text{upper}}, \; y_{\text{lower}}$: expressed in terms of the variable x.

$y_{\text{upper}} - y_{\text{lower}}$: represents the height of a typical *vertical* rectangle.

(R3) If $f(x) = 0$ for all x in $[a, b]$, then R is the region bounded above by $y = g(x)$, below by $y = 0$, on the left by $x = a$, and on the right by $x = b$.

area of $R = \int_a^b (g(x) - 0)\, dx = \int_a^b g(x)\, dx$

(R4) $\int_c^d (g(y) - f(y))\, dy = \int_c^d (x_{\text{right}} - x_{\text{left}})\, dy$

$x_{\text{right}}, \; x_{\text{left}}$: expressed in terms of the variable y.

$x_{\text{right}} - x_{\text{left}}$: represents the height of a typical *horizontal* rectangle.

(R5) The area of a region R may be expressed both ways (integrating in x and integrating in y). But an antiderivative may be easier to find for one of the integrals.

(R6) The area between the curves $y = g(x)$ and $y = f(x)$ bounded on the left by $x = a$ and on the right by $x = b$ is

$$\text{Area between curves} = \int_a^b |g(x) - f(x)|\, dx$$

When computing the area between the curves, split the integral into a sum of integrals at each x in (a, b) where $g(x) - f(x)$ changes sign. $\triangle$

Example: Find the area of the region bounded by $y = x^2$, $y = x^3$, $x = -1$, and $x = 0$.

□

Example: Find the area of the region bounded by $y = \sin x$, $y = x$, $x = -\pi$, and $x = \pi$.

□

Example: Find the area of the region bounded by $y = e^x$, $y = -x + 1$, $x = 0$ and $x = 1$.

□

Example: Find the area of the region bounded by $y = -x + 4$, $y = -4x + 4$, and $y = 0$.

□

Example: Find the area of the region bounded by $y^2 = x$ and $y = x - 2$.

$\square$

Example: Find the area of the region bounded by $y = \sin x$, $y = \cos x$, $x = 0$, and $x = 2\pi$.

$\square$

5.6 Chapter summary

Chapter 5 is all about the integral. The derivative and the integral are the two most important ideas in calculus. The fundamental theorem of calculus relates the two concepts.

(S1) The integral $\int_a^b f(x)\,dx$ measures signed area.

(S2) The area function A_f is given by

$$A_f(x) = \int_a^x f(t)\,dt = \text{signed area bounded by } f \text{ from } a \text{ to } x.$$

Some examples suggest that the area function is an antiderivative of f. This idea is really an easy form of the fundamental theorem of calculus.

(S3) The fundamental theorem of calculus.

(1) Let f be a continuous function, defined on an open interval I containing a. The function A_f with rule

$$A_f(x) = \int_a^x f(t)\,dt$$

is defined for every x in I, and $\dfrac{d}{dx}(A_f(x)) = f(x)$.

(2) Let f be continuous on $[a, b]$, and let F be *any* antiderivative of f. Then

$$\int_a^b f(x)\,dx = F(b) - F(a).$$

(S4) The integral may be precisely defined as a limit of approximating sums. Although the algebra is tedious, approximating sums make good graphical sense. Use computers and calculators to find approximating sums.

(S5) There are lots of applications involving the integral.

(A1) If $f(t)$ is the rate at which a quantity varies at time t, then $\int_a^b f(t)\,dt$ is the amount by which the same quantity varies over the interval from $t = a$ to $t = b$.

(A2) The integral may be used to find the exact area bounded by curves in the plane.

Chapter 6

Finding Antiderivatives

6.1 Antiderivatives: the idea

Introduction:

(I1) This chapter is all about finding antiderivatives.
Given f, can you find a function F so that $F' = f$.

(I2) The FTC is good motivation for doing this.
If we can find an antiderivative, the definite integral is easy.

$$\int_a^b f(x)\,dx = F(b) - F(a)$$

Example: Find $\displaystyle\int_0^2 (e^x - x)\,dx$

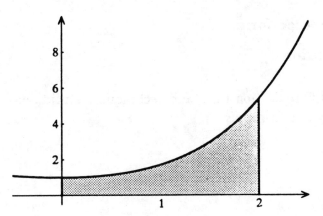

Finding antiderivatives

To use the FTC, we have to find an antiderivative. How do we know if there is one?

(F1) If f is continuous, then f has an antiderivative (FTC, part 1). The area function is an antiderivative

$$A_f(x) = \int_a^x f(t)\, dt$$

The problem is that the antiderivative might not have a nice formula, it might not be in closed form.

(F2) Remember, if a function has an antiderivative, it has infinitely many. Any two antiderivatives differ by a constant. So, finding one antiderivative is the same as finding them all.

Remarks:

(R1) Every elementary function has a derivative. But, finding antiderivatives is different (and harder).

An antiderivative of an elementary function may not be elementary. And, for some functions, deciding if an elementary antiderivative exists, and finding one, is very difficult.

(R2) So, here is the problem for the entire chapter:

Given an elementary function f, find an elementary function F such that $F' = f$.

The problem is called **integration in closed form**.

A solution is called a **closed form solution**.

(R3) So, we will have to develop some antidifferentiation rules and techniques. Antidifferentiation takes a lot of practice, and good *integral vision*.   △

Notation and terminology

(N1) Consider the expression $\displaystyle\int f(x)\, dx$

Called an **indefinite integral**.
f is the **integrand**.
x is the **variable of integration**.

(N2) In $\int x\,dx = \dfrac{x^2}{2} + C$ C is the **constant of integration**.

The definite integral represents a family of all possible antiderivatives of f.

(N3) Consider the expression $\int_a^b f(x)\,dx$

Called a **definite integral**.

Note!

(N1) The FTC links definite and indefinite integrals.

In words: The definite integral $\int_a^b f(x)\,dx$ equals the change $F(b) - F(a)$ from $x = a$ to $x = b$ in any indefinite integral.

(N2) Numerical methods let us approximate many definite integrals, without actually finding an antiderivative. That's a good thing, because in the real world there are lots of elementary functions without formulas or nice antiderivatives.

(N3) Be careful when using your calculator and the computer.

$$\int_1^2 \frac{1}{x}\,dx = \ln 2 \neq .6931471806$$

(N4) The dx inside the integral means something: x is the variable of integration, every other letter is a constant!

$$\int kx^2\,dx =$$

$$\int qe^x\,dx =$$

(N5) So, antidifferentiation is hard!
But, you can always check your answer by taking the derivative. $\diamond$

The next two examples are designed to help us start thinking about some antidifferentiation rules.

Example: $\int x e^x \, dx =$

□

Example: $\int \frac{x}{e^x} \, dx =$

□

Basic antiderivative formulas

$$\int x^k \, dx = \frac{x^{k+1}}{k+1} + C \quad (k \neq -1) \qquad \int \frac{1}{x} \, dx = \ln |x| + C$$

$$\int e^x \, dx = e^x + C \qquad \int a^x \, dx = \frac{a^x}{\ln a} + C \quad (a \neq 1)$$

$$\int \sin x \, dx = -\cos x + C \qquad \int \cos x \, dx = \sin x + C$$

$$\int \sec^2 x \, dx = \tan x + C$$

$$\int \frac{dx}{\sqrt{1-x^2}} \, dx = \arcsin x + C \qquad \int \frac{dx}{1+x^2} \, dx = \arctan x + C$$

Note!

(N1) Why are these formulas true?
Check by taking the derivative.

(N2) Why is there an absolute value symbol in the second formula?
The answer has to do with domains.
An antiderivative should have the same domain as the integrand. ◇

Two rules

(R1) $\displaystyle\int (f(x) + g(x)) \, dx = \int f(x) \, dx + \int g(x) \, dx$

In words: The integral of a sum is the sum of the integrals.

(R2) $\displaystyle\int k \cdot f(x) \, dx = k \cdot \int f(x) \, dx$

Constants pass freely through integral signs.

Example: $\int \left(5x^2 - 2x^{-3}\right) dx =$

$\square$

Example: $\int \dfrac{2x^4 - 3x^3 - x^2 + 5}{x^2}\, dx =$

$\square$

Example: $\int \left(\sin x + \sec^2 x + e^x\right) dx =$

$\square$

Example: $\int e^{-3x}\, dx =$

□

Example: $\int \sin x \cos^4 x\, dx =$

□

6.2   Antidifferentiation by substitution

The most important technique of integration is called **direct substitution**, u **substitution**, **change of variable**, or just plain **substitution**. It is like a "backwards chain rule." We will look at a general procedure, and then do lots of examples.

Remarks:

(R1) To use the substitution rule, you must "bring everything into the u world" where the problem is (hopefully) easier.

(R2) Think of dx and du as quantities obtained from the change of variable from x to u via $u = g(x)$:

$$\int f(g(x))g'(x)\, dx = \int f(u)\, du \qquad \begin{aligned} u &= g(x) \\ du &= g'(x)\, dx \end{aligned}$$

(R3) A Procedure:

   (S1) Select a substitution $u = g(x)$

   (S2) Find the antiderivative in terms of u :

$$\int f(g(x))g'(x)\, dx = \int f(u)\, du = F(u) + C$$

   (S3) Substitute $u = g(x)$ to express the antiderivative in terms of x, the original variable:

$$\int f(g(x))g'(x)\, dx = F(g(x)) + C$$

Example: $\displaystyle \int x(x^2 + 1)^4\, dx =$

Example: $\displaystyle \int x \sin x^2 \, dx =$

$\square$

Example: $\displaystyle \int \cos^3 x \sin x \, dx =$

$\square$

Example: $\displaystyle \int \frac{dx}{x^2 + 2x + 1} =$

$\square$

Example: $\displaystyle\int \frac{x^2}{\sqrt{x+1}}\, dx =$

□

Example: $\displaystyle\int \frac{x+1}{x^2+1}\, dx =$

□

Example: $\int \dfrac{(\ln x)^2}{x}\, dx =$

□

Example: $\int \dfrac{\cos x}{1 - \sin x}\, dx =$

□

Example: $\int \dfrac{1}{9 + x^2}\, dx =$

□

Theorem 1. Let f, u, and g be continuous functions such that for all x in $[a, b]$,

$$f(x) = g(u(x)) \cdot u'(x).$$

Then

$$\int_a^b f(x)\, dx = \int_{u(a)}^{u(b)} g(u)\, du.$$

Example: $\displaystyle\int_0^3 e^{-x^2} x\, dx =$

□

Example: $\displaystyle\int_1^3 (x^2 + 1)^2 x\, dx =$

□

Example: $\displaystyle\int_0^{\sqrt{\pi/3}} \sec^2(x^2)x\,dx =$

□

Example: $\displaystyle\int x^2 \sqrt[3]{2x+1}\,dx =$

□

Note!

(N1) Integrals for symmetric functions: Suppose f is continuous on $[-a, a]$.

    (S1) If f is even $[f(-x) = f(x)]$, then $\displaystyle\int_{-a}^{a} f(x)\,dx = 2\int_{0}^{a} f(x)\,dx$.

    (S2) If f is odd $[f(-x) = -f(x)]$, then $\displaystyle\int_{-a}^{a} f(x)\,dx = 0$.

        Example: $\displaystyle\int_{-4}^{4} \frac{x\cos x}{x^2 + 1}\,dx =$

$\square$

(N2) The symbols du and dx are called the **differentials** of u and x, respectively.

    If $u = u(x)$, then $du = u'(x)\,dx$

    For us, the differential in an integral expression indicates the variable of integration.$\diamond$

6.3 Integral aids: tables and computers

Finding symbolic antiderivatives and/or the "right" substitution can be difficult. However, the results can be pleasing and some important patterns may be revealed.

In practice, we often use integral tables or computer software to solve integral problems. There is a short table of integrals at the end of the book. Sometimes (most of the time) a given integral will require a transformation in order to correspond with one of the forms in the table of integrals.

Fact: (Formula 21)
$$\int \frac{x}{ax+b}\, dx = \frac{x}{a} - \frac{b}{a^2} \ln|ax+b|$$

Note!

(N1) There is a missing but *understood* constant of integration.

(N2) a and b are **parameters**, constants, letters other than the variable of integration. ◇

Example: Verify Formula 21 by differentiation.

□

Example: Use Formula 21 to find $\displaystyle\int \frac{\sin x \cos x}{1 - \sin x}\, dx$

Example: $\displaystyle\int e^{\sqrt{x}}\, dx =$

□

□

Example: $\displaystyle\int \frac{dx}{\sqrt{x^2 + 4x + 13}} =$

□

Example: $\displaystyle\int e^{2x} \sin(-3e^x)\, dx =$

□

Note! The symbolic antiderivatives given in integral tables are a little mysterious: Where did those answers in the back of the book come from? Do we need to verify these formulas by differentiation?

Remember: An integrand f and an antiderivative F must *agree graphically*. You can always symbolically check your answer: Does $F'(x) = f(x)$? But the geometry of derivatives must also hold and serves as a graphical check of your answer.   ◇

Some formulas in an integral table have antiderivatives on *both* sides of an equation. Here is an example of a **reduction formula**:

For n an integer and $n \geq 2$

$$\int \sin^n(ax)\,dx = -\frac{\sin^{n-1}(ax)\cos(ax)}{na} + \frac{n-1}{n}\int \sin^{n-2}(ax)\,dx$$

Example: $\int \sin^4(-3x)\,dx =$

Remarks:

(R1) Mathematical software now produces symbolic antiderivatives. Programs like *Mathematica*, *Maple*, and *Derive* have replaced integral tables.

Sometimes two computer programs will return *different looking* antiderivatives, and both may be different from the one produced *by hand*.

Recall: Two antiderivatives must differ by a constant. Sometimes trigonometric identities or fancy algebra is necessary to verify this fact.

(R2) Occasionally symbolic methods, integral tables, algebraic tricks, and even software all fail to find an antiderivative in elementary form.

(F1) It could be that we haven't used the right algebraic trick or transformation, and the computer can't *see* it either.

(F2) Or, there may be no elementary antiderivative.

Remember, a lot of very simple elementary functions do not have elementary antiderivatives.

(R3) If there is no symbolic antiderivative, numerical methods will at least treat *definite* integrals of troublesome functions. △